T0140048

Springer Theses

Recognizing Outstanding Ph.D. Research

Aims and Scope

The series "Springer Theses" brings together a selection of the very best Ph.D. theses from around the world and across the physical sciences. Nominated and endorsed by two recognized specialists, each published volume has been selected for its scientific excellence and the high impact of its contents for the pertinent field of research. For greater accessibility to non-specialists, the published versions include an extended introduction, as well as a foreword by the student's supervisor explaining the special relevance of the work for the field. As a whole, the series will provide a valuable resource both for newcomers to the research fields described, and for other scientists seeking detailed background information on special questions. Finally, it provides an accredited documentation of the valuable contributions made by today's younger generation of scientists.

Theses are accepted into the series by invited nomination only and must fulfill all of the following criteria

- They must be written in good English.
- The topic should fall within the confines of Chemistry, Physics, Earth Sciences, Engineering and related interdisciplinary fields such as Materials, Nanoscience, Chemical Engineering, Complex Systems and Biophysics.
- The work reported in the thesis must represent a significant scientific advance.
- If the thesis includes previously published material, permission to reproduce this must be gained from the respective copyright holder.
- They must have been examined and passed during the 12 months prior to nomination.
- Each thesis should include a foreword by the supervisor outlining the significance of its content.
- The theses should have a clearly defined structure including an introduction accessible to scientists not expert in that particular field.

More information about this series at http://www.springer.com/series/8790

Amandeep Kaur

Fluorescent Tools for Imaging Oxidative Stress in Biology

Doctoral Thesis accepted by
the University of Sydney, Australia

 Springer

Author
Dr. Amandeep Kaur
School of Chemistry
University of Sydney
Sydney, NSW
Australia

Supervisor
Dr. Elizabeth J. New
School of Chemistry
University of Sydney
Sydney, NSW
Australia

ISSN 2190-5053   ISSN 2190-5061   (electronic)
Springer Theses
ISBN 978-3-030-10371-2   ISBN 978-3-319-73405-7   (eBook)
https://doi.org/10.1007/978-3-319-73405-7

Printed on acid-free paper

This Springer imprint is published by Springer Nature
The registered company is Springer International Publishing AG
The registered company address is: Gewerbestrasse 11, 6330 Cham, Switzerland

Supervisor's Foreword

The ability to visualise cellular processes in real time is crucial to understand disease development and streamline treatment. This can be achieved by fluorescent chemical tools that can reversibly sense disturbances in cellular environments during pathogenesis. In particular, perturbations in cellular redox state are of current interest in medical research, with oxidative stress now implicated in pathogenesis of many diseases.

The work described here addressed the lack of tools to study redox changes within cells with temporal resolution: despite the plethora of reaction-based probes for reactive oxygen species, there were very few reversible sensors. Importantly, this work represents the first comprehensive studies of fluorescent sensors based on the endogenous flavin redox switch. Strategies were also investigated and applied to achieve ratiometric fluorescence output of the reversible redox probes, which nullify concentration effects associated with intensity-based probes.

The work also describes suitable strategies to target these probes to specific cellular organelles, thereby enabling medical researchers to visualise subcellular oxidative stress levels and addressing the typically poor uptake of chemical tools into biological studies. In total, four new redox probes are reported in this thesis, which have already found use in over twenty research groups worldwide. The final chapters of the thesis demonstrate successful applications of the sensors in a variety of biological systems ranging from prokaryotes to mammalian cells, to whole organisms, highlighting the application of the developed small-molecule chemical sensors in the diagnosis of diverse pathophysiological conditions.

Significant progress in the field of chemical biology requires genuine, cross-disciplinary collaboration, to drive development of the most useful tools and to ensure their impactful application. This work involved the direct input of six other research groups in Australia, with varied expertise from electrochemistry to immunology. In particular, the involvement of medical researchers, who will be the end users of the probes described herein, ensured that target design focussed on probes that are already having significant impact in the understanding of health and disease.

Sydney, Australia   Dr. Elizabeth J. New
September 2017

Abstract

Oxidative stress has been implicated in a myriad of pathological conditions, but measuring it and deciphering the underlying mechanisms has been a long-standing challenge. Most of the existing redox probes are reaction-based and irreversible. In contrast, reversible redox probes with biologically tuned redox potential can distinguish between transient and chronic elevations in oxidative capacity.

In this work, a series of novel fluorescent redox probes based on flavins and nicotinamides have been developed and tested. The first redox-responsive probe, **NpFR1**, localises in the cytoplasm and exhibits robust reversibility of oxidation and reduction and a clear fluorescence increase upon oxidation. Based on these promising results, attention turned to developing analogues of **NpFR1**, with two main aims: to send the probe to the specific subcellular organelles and to develop a ratiometric probe, in which a change in redox state was signalled by a change in colour rather than fluorescence intensity.

To this end, six further probes were developed. A mitochondrially localising derivative **NpFR2** was developed by incorporating a lipophilic cationic tag. The design of the emission-ratiometric redox probe **FCR1**, whose fluorescence emission changes from blue to green upon oxidation, is based on the FRET process between coumarin donor and flavin acceptor. Furthermore, a set of excitation-ratiometric FRET probes **FRR1** and **FRR2** were developed, in which mitochondrial targeting was achieved by employing rhodamine B with inherent mitochondrial accumulation properties. Finally, nicotinamide-based redox probes **NCR3** and **NCR4** demonstrate the use of ICT as a strategy to attain ratiometric fluorescence properties.

Following the successful utilisation of ratiometric and targeting strategies, the developed probes were applied in a variety of biological investigations, whether *in cellulo*, *ex vivo* or non-mammalian systems. The probes demonstrated excellent abilities to report on the differences in oxidative capacities under different conditions within each system. This work therefore demonstrates that reversible redox probes based on flavins and nicotinamides exhibit suitable properties for use as cellular redox probes. The developed probes can be modulated to give ratiometric output and targeted to specific subcellular compartments. These probes therefore exhibit potential to aid in deciphering the role of oxidative stress in pathogenesis and disease progression.

Parts of this thesis have been published in the following journal articles:

1. Jonathan Yeow, Amandeep Kaur, Matthew D. Anscomb and Elizabeth J. New, 2014, A novel flavin derivative reveals the impact of glucose on oxidative stress in adipocytes. *Chemical Communications*, **50**, 8181–8184. **DOI:** 10.1039/C4CC03464C
2. Amandeep Kaur, Mohammad A. Haghighatbin, Conor F. Hogan and Elizabeth J. New, 2015, A FRET-based ratiometric redox probe for detecting oxidative stress by confocal microscopy, FLIM and flow cytometry. *Chemical Communications*, **52**, 10510–10513. **DOI:** 10.1039/C5CC03394B
3. Amandeep Kaur, Kurt W. L. Brigden, Timothy F. Cashman, Stuart T. Fraser and Elizabeth J. New, 2015, Mitochondrially targeted redox probe reveals the variations in oxidative capacity of the haematopoietic cells. *Organic & Biomolecular Chemistry*, **24**, 6686–6689. **DOI:** 10.1039/C5OB00928F
4. Amandeep Kaur, Jacek L. Kolanowski and Elizabeth J. New, 2016, Reversible fluorescent probes of biological redox state. *Angewandte Chemie International Edition*, **55**, 1602–1613. **DOI:** 10.1002/anie.201506353
5. Jacek L. Kolanowski, Amandeep Kaur and Elizabeth J. New, 2016, Selective and reversible approaches towards imaging redox signalling using small molecule probes. *Antioxidants and Redox Signaling*. **DOI:** 10.1089/ars.2015.6588
6. Amandeep Kaur, Karolina Jankowska, Chelsea Pilgrim, Stuart T. Fraser and Elizabeth J. New, 2016, Studies of haematopoietic cell differentiation with a ratiometric and reversible sensor of mitochondrial oxidative capacity. *Antioxidants and Redox Signaling*. **DOI:** 10.1089/ars.2015.6495

Acknowledgements

First and foremost I thank my supervisor Dr. Elizabeth J. New; it has been an honour to be your first Ph.D. student. Your unlimited support and enthusiasm made my Ph.D. experience some of the best years of my life. The passion you have for this research was so contagious and motivational, and helped me smoothly navigate the Ph.D. roller-coaster. Liz, you have been an irreplaceable mentor for me—I couldn't have asked for a better one! I appreciate the encouragement and opportunities you have given me to grow as a scientist so much over the past three years, and for being someone I can always talk to about anything. Your kindness and confidence will no doubt be an enduring quality that I will cherish for all times to come.

I owe endless thanks to the outstanding members of the New Group for offering friendly advice, assistance with experiments and proofreading, and for the amazing social life I had during my Ph.D. Most importantly, the bonding and chats over endless group lunches, and the competitive bowling and games nights. Thank you Clara and Karolina for making my weekends enjoyable, whether having brekkies or shopping, not to mention for getting me to love the idea of noodles in soup! Clara, special thanks to you for being my encyclopaedia, walking around with answers to almost every question I had. Thanks Ed for helping me with computer issues. I also thank the postdocs David, Jacek, Ze and Mia for their unfailing guidance throughout. David, thank you for stimulating my interest in bowling and for chatting with Matt 2.0! I do not have a preference for rice over noodles! Jacek, thank you for looking after the cells when I have been away, for your pep talks and for bearing my silly 'no' for an answer. Also Jacek, yes, it is 50 μM peroxide for 30 minutes! I would also like to extend my thanks to Ze and Mia for their extraordinary guidance with job applications. Isaac, thanks so much for being the one person who understood cricket, and for explaining me the content on certain card games. I am also thankful for the advice and company of Angela, Kate, Linda, Marcus, Lucy and Laura. I thank the former Honours students of the New Group Matt, Elliot and Jono for helping me settle in when I joined the lab. Maddy, you bake amazing cupcakes, Kylie, thank you for having the same love for dogs as I do.

Carmen, thanks for doing a TSP with me, you will forever remain the first student I ever supervised.

The multidisciplinary nature of this research was made possible by the unmatched assistance and guidance of our collaborators. I am grateful to Dr. Stuart Fraser, Dr. Gawain McColl, Assoc. Prof. Alexandra Sharland, Dr. Andrew Hoy, Dr. Anahit Penesyan, Dr. Michael Cahill and the members of their research labs for the knowledge and training and for the opportunity to work with various biological systems. I am also grateful to Assoc. Prof. Conor Hogan and his group for hosting me and for help with spectro-electrochemical studies.

The technical and professional staff in the School Ian—Nick, Jeff, Fernando, Ellen, Dimetra and Jody—have been of great help throughout my Ph.D. I would also like to thank the ACMM staff Ellie, Minh, Pam and Su for their help with cell culture and microscopy. Thanks to Adrian and Steven at the Centenary Institute for help with flow cytometry, and to the technical staff at BOSCH—Angeles for help with flow cytometry, and Donna and Sheng for assistance with Seahorse assays.

I would also like to extend my thanks to Assoc. Prof. Ron Clarke, Assoc. Prof. Paul Whitting, Dr. Aviva Levina and Mr. Jun Liang for graciously assisting me and sharing their ideas and expertise. I am grateful to the John A. Lamberton and the University of Sydney World scholars scholarship for financial assistance during my Ph.D.

Finally, I would like to thank friends and family for their understanding and many good times. Special thanks to my Mum who supported, encouraged and prayed for me over the years. Thanks Mum for sticking by me, no matter where I was.

Contents

Abbreviations

°C	Degrees Celsius
μM	Micromolar
μm	Micrometre
λ_{ex}	Emission wavelength
λ_{em}	Excitation wavelength
AA	Antibiotic/antimycotic
AB	*Acinetobacter baumannii*
AGE	Advanced glycation end products
AIE	Aggregation-induced emission
ALD	Alcoholic liver disease
AO	Antioxidants
APCI	Atmospheric pressure chemical ionisation
ARDS	Adult respiratory distress syndrome
ATP	Adenosine triphosphate
BC	*Bacillus cereus*
Bcl-2	B-cell lymphoma 2
BODIPY	Boron-dipyrromethene
BSA	Bovine serum albumin
BSO	Buthionine sulfoximine
CAT	Catalase
CTR1	Copper transporter 1
CuAAC	Copper-catalysed azide-alkyne cycloaddition
DCC	*N,N'*-Dicyclohexylcarbodiimide
DCM	Dichloromethane
DFT	Density functional theory
DIPEA	*N,N'*-Diisopropylethylamine
DM	Double mutant
DMEM	Dulbecco's modified eagle medium
DMF	*N,N*-Dimethyl formamide
DMSO	Dimethyl sulfoxide

dpc	*Days post coitum*
DTT	Dithiothreitol
ECAR	Extracellular acidification rate
ELISA	Enzyme-linked immunosorbent assay
EryD	Definitive erythroid
EryP	Primitive erythroid
ESCs	Embryonic stem cells
ESI	Electron spray ionisation
ESIPT	Excited-state induced proton transfer
ESR	Electron spin resonance
EtOAc	Ethyl acetate
EtOH	Ethanol
FAD	Flavin adenine dinucleotide
FBS	Foetal bovine serum
FCCP	Carbonyl cyanide-4-(trifluoromethoxy)phenylhydrazone
FCS	Foetal calf serum
FLIM	Fluorescence lifetime imaging microscopy
FMN	Flavin mononucleotide
FRET	Forster resonance energy transfer
GFP	Green fluorescent protein
GI	Gastrointestinal
GPXs	Glutathione peroxidases
GRX	Glutaredoxin
GSH	Glutathione
GSH/GSSG	Ratio of reduced and oxidised glutathione
HEPES	(4-(2-Hydroxyethyl)-1-piperazineethanesulfonic acid)
HNE	4-Hydroxy-2-nonenal
HPLC	High-performance liquid chromatography
HSCs	Haematopoietic stem cells
ICT	Intramolccular charge transfer
LB	Lysogeny broth
LPS	Lipopolysaccharide
MDA	Malondialdehyde
MeCN	Acetonitrile
MeOH	Methanol
MHz	Megahertz
MP	MiaPaCa-2
MTT	(3-(4,5-Dimethylthiazol-2-yl)-2,5-diphenyltetrazolium bromide
mV	Millivolts
NAC	N-acetylcysteine
NAD	Nicotinamide adenine dinucleotide
NADP	Nicotinamide adenine dinucleotide phophate
NAFLD	Non-alcoholic fatty liver disease
NHS	*N*-Hydroxysuccinimide
NK	Natural killer cells

nm	Nanometer
nM	Nanomolar
NSAID	Non-steroidal anti-inflammatory drugs
OCR	Oxygen consumption rate
PA	*Pseudomonas aeruginosa*
PBS	Phosphate-buffered saline
PCC	Pearson's colocalisation coefficient
PET	Photo-induced electron transfer
PGRMC1	Progesterone receptor membrane component 1
PI	Propidium iodide
PMA	Phorbol-12-myristate-13-acetate
PPC	Particles per cell
ppm	Parts per million
PRXs	Peroxiredoxins
PUFAs	Polyunsaturated fatty acids
RNA	Ribonucleic acid
RNS	Reactive nitrogen species
ROS	Reactive oxygen species
RT	Room temperature
S/N	Signal-to-noise ratio
SHE	Standard hydrogen electrode
SOD	Superoxide dismutase
TBAB	Tetrabutylammonium bromide
TBAPF6	Tetrabutylammonium hexafluorophosphate
TCSPC	Time-correlated single photon counting
TLC	Thin-layer chromatography
TLR4	Toll-like receptor 4
TM	Triple mutant
TPP	Triphenylphosphonium
TRX	Thioredoxin
U.V.	Ultraviolet
V	Volts
Vs^{-1}	Volts per second
WT	Wild-type

List of Figures

List of Schemes

List of Tables

Chapter 1
Introduction

1.1 The Toxicity of Oxygen

Primordial life forms on earth comprised oxygen-sensitive organisms: the anaerobic fermenters and cyanobacteria, which released oxygen as a metabolic by-product, causing the oxygen levels in the atmosphere to rise [1, 2]. Consequently, aerobic respiration processes such as photosynthesis and O_2-dependent electron transport mechanisms, came to predominate in biological systems [1, 2]. The paradox of aerobic respiration is that oxygen, which is vital for life is also a fatal toxin. The consequence of aerobic respiration is the generation of partially-reduced forms of oxygen—reactive oxygen species (ROS), which oxidatively damage important biomolecules such as DNA, lipids and proteins [3, 4]. As a result, highly coordinated and effective antioxidant (AO) defence mechanisms evolved, which counteract and neutralise the toxic ROS [1, 2]. Although well integrated, the AO defences are not infallible [5].

The toxic effects of ROS were first established in 1954 by Gerschman et al. [6] and since then a large volume of evidence has accumulated that demonstrate the role of ROS in a myriad of pathological conditions [7–9]. However, more recently, the role of ROS as signalling molecules with a crucial role in orchestrating specific physiological functions has been realised [5, 10, 11]. Understanding the importance of ROS and AO in health and disease has therefore been an intense area of scientific research. However, the current challenge faced by researchers studying cellular redox state is the design of a robust method that enables real time measurement of redox state in cells [12, 13]. This chapter contains a review of the factors that determine the redox state of a cell and its importance. The advantages and disadvantages of the various tools currently used to study the redox state are discussed with particular focus on fluorescence imaging techniques.

Parts of the text and figures of this chapter are reprinted with permission from *Antioxidants and Redox Signalling*, Volume 24, Issue 13, published by Mary Ann Leibert, Inc., New Rochelle, NY., and *Angewandte Chemie*, Volume 55, Issue 5, published by John Wiley & sons, Inc.

1.1.1 Reactive Oxygen and Nitrogen Species (ROS/RNS)

Most eukaryotic biological systems carry out aerobic respiration in the mitochondria to generate ATP, the energy currency of the cell. This cellular oxidative metabolism is driven by an electron transport reaction, in which O_2 accepts electrons and H^+, and is eventually reduced to water. Although the mitochondrial electron transport chain is extraordinarily efficient, 1–3% of all electrons involved can leak out and be transferred to O_2, resulting in the formation of the superoxide anion radical ($O_2^{•-}$, Eq. 1.1), the primary reactive oxygen species (ROS) [14]. Superoxide can also be produced enzymatically by NADPH oxidase [14].

$$O_2 + e^- \rightarrow O_2^{•-} \tag{1.1}$$

Superoxide is a moderately reactive, short-lived ROS (half-life $\simeq 1$ μs) and therefore not believed to cause direct oxidation of cellular components (DNA, protein and lipids). Instead it produces secondary ROS and reactive nitrogen species (RNS) (Fig. 1.1), whether directly or via enzyme- and metal-catalysed processes [5]. Secondary ROS include hydrogen peroxide (H_2O_2), the hydroxyl radical ($^•OH$), hypochlorite ion (OCl^-) and singlet oxygen ($^1O_2$), whereas RNS include the nitric oxide radical ($NO^•$) and the peroxynitrite anion ($ONOO^-$) [3].

Superoxide rapidly undergoes dismutation, catalysed by a family of enzymes called superoxide dismutases (SODs), to form H_2O_2 (Eq. 1.2, Fig. 1.1), which is relatively more stable (half-life $\simeq 1$ ms) [16, 17]. Unlike other ROS, H_2O_2 is a

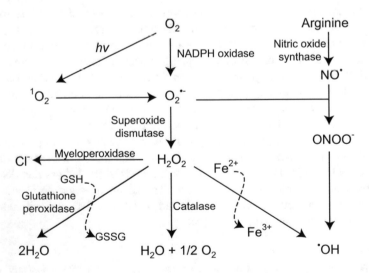

Fig. 1.1 An overview of the production of primary and secondary ROS/RNS. Antioxidants are shown in blue [4, 14, 15]

neutral molecule capable of diffusing across biological membranes. This property allows H_2O_2 to act as a signalling molecule, but at high concentrations it can cause oxidative damage over large distances within the cell [16].

$$2O_2^{\bullet -} + 2H^+ \xrightarrow{SOD} H_2O_2 + O_2 \qquad (1.2)$$

Low levels of H_2O_2 are maintained by the action of a family of enzymes—glutathione peroxidases (GPXs), which transfer electrons from two molecules of glutathione (GSH) to H_2O_2 to form water (Eq. 1.3, Fig. 1.1) [16]. The enzyme catalase can convert six million H_2O_2 molecules per minute to water and oxygen (Eq. 1.4) [16].

$$2GSH + H_2O_2 \xrightarrow{GPX} GSSG + H_2O \qquad (1.3)$$

$$H_2O_2 \xrightarrow{Catalase} H_2O + O_2 \qquad (1.4)$$

The hydroxyl radical ($^\bullet OH$) is formed by Haber-Weiss (Eq. 1.5) catalysed by redox active metals such as Fe and Cu (Fig. 1.1). It is the most reactive among all ROS, tending to react very close to its site of formation. The redox active metals Fe and Cu are closely associated with anionic species such as cell membranes and DNA, hence $^\bullet OH$-mediated oxidative damage occurs in close proximity to these species [18]. Furthermore, because the cells lack specific mechanisms to eliminate $^\bullet OH$, its excess production can have severe pathological consequences [19].

$$O_2^{\bullet -} + H_2O_2 \xrightarrow{Fe^{3+}} {}^\bullet OH + {}^- OH \qquad (1.5)$$

The hypochlorite ion (OCl^-), is formed from the reaction of H_2O_2 with chloride ions (Cl^-), catalysed by the enzyme myeloperoxidase (MPO, Eq. 1.6, Fig. 1.1) [20]. The myeloperoxidase-H_2O_2-Cl^- system is crucial for the proper functioning of phagocytes, giving these cells the ability to kill a wide range of pathogens. Although OCl^- is essential for a proper immune response, uncontrolled production of HOCl within the phagocytes is detrimental to the host tissue [20].

$$H_2O_2 + Cl^- \xrightarrow{MPO} OCl^- + {}^\bullet OH + Fe^{3+} \qquad (1.6)$$

Singlet oxygen ($^1O_2$), unlike other ROS, results from the UV illumination of molecular oxygen (Fig. 1.1), and is moderately stable (half-life $\simeq 3\mu s$). At high concentrations, it diffuses across the nuclear membrane and oxidatively damages the DNA through selective oxidation of guanine to 8-hydroxylguanine. However, $^1O_2$ can be quenched by vitamins A (β-carotene) and E (α-tocopherol).

The primary source of RNS is the nitric oxide radical ($NO^\bullet$), which is synthesised endogenously from L-arginine by the enzyme nitric oxide synthase (Fig. 1.1). Following the Nobel prize-winning discovery of nitric oxide as a signalling molecule, the physiological significance of this gaseous radical species has gained great attention

Table 1.1 The sources of ROS

Endogenous sources	**Electron transport chain**—mitochondria, microsomes and chloroplasts.
	Oxidant enzymes—lipoxygenase, monoamine oxidase, indoleamine dioxygenase, tryptophan dioxygenase, xanthine oxidase, galactose oxidase.
	Phagocytic cells—endothelial cells, neutrophils, macrophages, monocytes, eosinophils.
	Autooxidation reactions—adrenalin, thiols, Fe^{2+}
Exogenous sources	**Redox cycling substances**—substances that are reduced in vivo to a radical that then react with oxygen to regenerate the drug and form a ROS for example, paraquat, alloxan, doxorubicin and diquat.
	Other—U.V. radiation, drug oxidation, cigarette smoke, ionising radiation, pollution and heat shock

[21, 22]. NO• acts as an effector molecule during immune response, as a neurotransmitter and a smooth muscle relaxing factor [23]. It is also implicated in pathological activity such as inflammation, neurodegenerative tissue injury and vasodilation [24]. Direct reaction of NO• with $O_2^{•-}$ results in the production of peroxynitrite (ONOO⁻, Eq. 1.7, Fig. 1.1) [25]. With a short half-life ($\simeq 1\,\mu$s), ONOO⁻ is highly reactive, oxidising tyrosine residues of proteins to form stable 3-nitrotyrosine thereby impacting the structure and functions of proteins [25, 26].

$$O_2^{•-} + NO^{•} \rightarrow ONOO^{-} \tag{1.7}$$

In addition to leakage from the electron transport chain, reactive oxygen and nitrogen species (ROS/RNS) can be produced in the cell from various other endogenous and exogenous sources (Table 1.1) [4, 15].

1.1.2 Scavenging ROS/RNS

Deleterious effects of ROS/RNS make their scavenging all the more essential, and have resulted in the evolution of a series of defence mechanisms mediated by cellular AO (Fig. 1.1). AO act in an altruistic manner, preferentially oxidising themselves in order to protect the cellular components. They can be classified into three groups based on their origin [27].

1. **Endogenous AO** are those that are synthesised naturally by the cells.

   • **Enzymatic** endogenous AO include superoxide dismutase (SOD), catalase (CAT), glutathione peroxidases (GPXs), thioredoxins(TRX) and peroxiredoxins(PRXs).

- **Non-enzymatic** endogenous AO are small molecules (L-carnitine, glutathione, α-lipoic acid, uric acid and bilirubin), coenzymes (coenzyme Q) and low molecular weight proteins (metallothionein, coenzyme Q, ferritin and melatonin) [27].

2. **Natural AO** are usually present in daily dietary intake like, ascorbic acid (Vitamin C), tocopherol (Vitamin E) and β-carotene (Vitamin A), lipoic acids, and polyphenol metabolites [27].
3. **Synthetic AO** are the active ingredients of AO supplements. Examples include N-acetyl cysteine (NAC), tiron, pyruvate, selenium, propyl gallate, butylated hydroxyltoluene and butylated hydroxylanisole [27].

The cellular repair apparatus comprising AO, DNA repair enzymes and proteases (that degrade damaged proteins) is tightly regulated and is essential in maintaining the cellular redox homoeostasis with low levels of ROS/RNS [28].

1.2 ROS/RNS—Friends and Foe

ROS/RNS are produced in biological systems by normal cellular metabolism. Substantial evidence has been documented that ROS/RNS play a dual role in biological systems—as both beneficial and deleterious species [10, 11, 29]. The levels of ROS/RNS naturally fluctuate in the cells, with well-regulated increases in the production of ROS/RNS ensuring uncompromised physiological functions. On the other hand, uncontrolled increases would result in pathogenesis. It is the AO that play a major role in maintaining healthy levels of these species by scavenging them.

1.2.1 ROS/RNS and Physiology

Transient elevations in the concentrations of ROS/RNS have proved to be essential for many sub-cellular functions ranging from gene expression, enzyme activation and signal transduction to folding of proteins in the endoplasmic reticulum, vascular homoeostasis and apoptosis [30–33]. A widely accepted mechanism of redox signalling involves the reversible oxidation of redox-sensitive thiol residues (in thiols or low molecular weight thioproteins), thereby modulating their function and activity. While the majority of cellular thiols exist in the protonated form (-SH) at physiological pH, a small fraction of them will remain in the form of thiolate anions (-S$^-$), which can also be favoured in a particular local protein microenvironment, increasing the susceptibility to oxidation when compared to the conjugate acid form (-SH) [34, 35]. The oxidation of the thiolate anion by H_2O_2, for example, results in the generation of a sulfenic acid (-SOH), causing allosteric modification of protein structure and function.

This sulfenic acid form enters the signal transduction pathway, in which it is reversibly reduced to the thiolate anion by glutaredoxin (GRX) and thioredoxin (TRX), restoring the original structure and function of the protein [36]. While transient increases in the generation of ROS/RNS bring about a reversible oxidation of thiolate anion to sulfenic acid, chronic elevations can result in further irreversible oxidations, generating sulfinic ($-SO_2H$) and sulfonic ($-SO_3H$) residues, rendering the protein structure and function irreparable, and thus resulting in oxidative stress [36]. Other beneficial effects of ROS/RNS include defence against infectious agents and induction of mitogenic response [30]. Concentrations of ROS/RNS higher than those required to perform such functions are immediately scavenged by AO, the cell's defence system.

1.2.2 ROS/RNS and Pathology

Owing to their highly reactive nature and the ability to oxidatively damage vital biomolecules, chronically elevated concentrations of ROS/RNS have been implicated in a variety of pathological conditions [29, 37, 38]. Evidence points to a role of ROS/RNS in the oxidation of purine and pyrimidine bases of DNA and RNA, especially the formation of 8-hydroxylguanine, as well as extensive damage to the deoxyribose backbone [15, 39]. Oxidation of DNA by these reactive species and subsequent genetic damage has been reported to induce aging, mutagenesis and carcinogenesis of cells [15].

ROS/RNS bring about a chain of peroxidative reactions of the polyunsaturated lipids in the biomembranes, consequently increasing the permeability of the cell membrane, leading to cell death. Peroxyl radicals ($ROO^{\bullet}$), in particular, have been reported to initiate a series of cyclisation reactions resulting in the formation of malondialdehyde (MDA), a potent carcinogen [4]. Functional properties of proteins and enzymes can be drastically altered or completely lost due to the oxidation-induced modification of protein structure by ROS/RNS [10]. Cysteine and methionine residues are the most vulnerable to oxidative damage, which mostly involves thiolation reactions. Thiolation reactions result in the reversible formation of disulfides and involve either intra-molecular reactions—with cysteine residues within the protein or intermolecular reactions with cysteine residues of other proteins or low molecular weight thiols like glutathione (GSH). Owing to the extensive damaging effects of ROS/RNS on vital biomolecules many physiological functions will be impaired, hindered or completely altered thereby resulting in a myriad of pathological conditions (Table 1.2) [5, 27, 40–46].

Transient elevations in the concentrations of ROS/RNS play a crucial role in signal transduction and related physiological functions following which the cellular AO levels are upregulated and the cellular redox homoeostasis is re-established. However, these repair mechanisms are prone to errors and mutations, and in conditions of chronic elevations in cellular oxidative capacity, the repair mechanisms fall short and the cell fails to return to its original state of redox homoeostases,

Table 1.2 Pathological conditions induced by ROS [5, 27, 40–46]

Organ/injury	Pathological conditions
Brain	Parkinson's disease, Lou-gehrig's disease, neurotoxins, Alzheimer's disease, multiple sclerosis, hypertensive cerebrovascular injury, aluminium overload, allergic encephalomyelitis, potentiation of traumatic injury
Eye	Photic retinopathy, ocular haemorrhage, cataractogenesis, retrolental fibroplasia, degenerative retinal damage, retinopathy of prematurity
Heart and cardiovascular system	Atherosclerosis, adriamycin cardiotoxicity, keshan disease, alcohol cardiomyopathy
Kidney	Metal ion mediated nephrotoxicity, aminoglycoside nephrotoxicity, Autoimmune nephrotic syndromes
Gastrointestinal tract	NSAID-induced GI tract lesions, oral iron poisoning, endotoxin liver injury, diabetogenic actions of alloxan, halogenated hydrocarbon liver necrosis, free fatty acid induced pancreatitis
Inflammatory immune injury	Rheumatoid arthritis, glomerulonephritis, autoimmune diseases, vasculitis (hepatitis B virus)
Iron overload	Nutritional deficiencies (kwashiorkar), thalassemia and other chronic anaemias, idiopathic haemochromatosis, Dietary iron overload
Red blood cells	Fanconi's anaemia, sickle cell anaemia, haemolytic anaemia, favism, Malaria, protoporphyrin photo-oxidation
Lungs	Bronchopulmonary dysplasia, mineral dust pneumoconiosis, bleomycin and paraquat toxicity, hypoxia, cigarette smoke effect (necrosis), emphysema, adult respiratory distress syndrome (ARDS)
Ischaemia reperfusion	Myocardial infarction/stroke, organ transplantation, frostbite
Other	Cancer, aging, alcoholism, radiation injury, porphyria, contact dermatitis

resulting in a heightened baseline oxidation state, a condition termed oxidative stress (Fig. 1.2). High local and global concentrations of ROS/RNS cause the damage of vital biomolecules like nucleic acids, lipids and proteins, and result in a myriad of pathological conditions.

1.2.3 Cellular Redox State

It is therefore evident that in order to maintain a healthy environment, cells are required to tune their redox balance, or oxidative capacity. Too low an oxidative

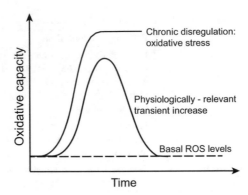

Fig. 1.2 Variations in cellular ROS levels. Transient increases are vital for physiological functions such as signalling, whereas chronic elevations are damaging and cause diseases. Reprinted with permission from *Angewandte Chemie*, Volume 55, Issue 5, published by John Wiley & sons, Inc.

capacity would mean sub-optimal signal transduction and other physiological functions, while too great an oxidative capacity is indicative of potentially-damaging ROS [47]. The oxidative damage caused by ROS depends on the intracellular concentrations of these reactive species as well as the equilibrium between ROS and AO. The redox balance of a cell, i.e. the pro-oxidant/anti-oxidant equilibrium, at any given time is representative of the cellular redox state, which is defined as the ability of a cell to act as a reducing agent or oxidising agent (Fig. 1.3). Lower concentrations of ROS or higher AO concentrations indicate a reducing cell environment. However, when the redox balance is perturbed in favour of ROS, beyond the cells ability to scavenge or repair the damage caused by ROS, oxidative stress develops, leading to oxidising cell environment [47].

Despite the established synergism between oxidative stress and disease, little is known about the extent of oxidative changes and the mechanisms involved. This clearly substantiates the need to develop tools that can report on the real time redox state of a cell and aid in elucidating the mechanisms involved.

1.3 Evaluating Cellular Redox State

To date, various methods have been developed and introduced to measure redox state in cells. However, many of these methods are not robust and suffer from drawbacks like simplicity, non-invasiveness, efficiency, and lack of real time monitoring of redox state. Traditionally, two different approaches have been exploited:

1. Measuring the consequences of ROS/RNS reactivity (biomarkers) in cellular fluids.
2. Directly quantifying ROS/RNS in cells.

Each will be addressed briefly below.

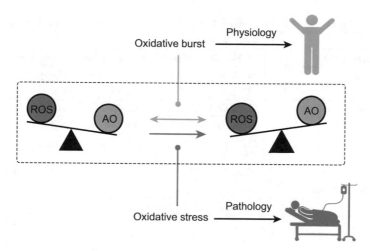

Fig. 1.3 The redox balance. Reversible increases in the concentrations of ROS are necessary for adequate cell physiology, whereas irreversible depletion of antioxidants (AO) or over production of ROS results in oxidative stress and related pathological consequences

1.3.1 Measuring Consequences of Oxidative Stress

Damaging effects of ROS/RNS on key biomolecules are well established (Sect. 1.2.2). This strategy focusses on quantifying the species produced as a result of the attack by ROS/RNS on various biomolecules. Several biomarkers of oxidative stress have been reported [40, 41, 48–56] as means of quantifying the damaging effects of ROS and establishing their role in inflicting a myriad of clinical conditions, these include:

1. Total antioxidant capacity of plasma.
2. Consumption of antioxidants.
3. Local activity of antioxidant enzymes.
4. In-vivo concentration of free radicals.
5. Markers of downstream consequences of oxidation.

Of all these, markers of oxidation have been extensively investigated, and are recognised hallmarks of various pathological conditions [55, 57–59]. As a result of these investigations, several key biomarkers associated with each disease have been identified (Table 1.3).

The above-mentioned biomarkers accumulate to detectable concentrations in cell, tissue or body fluids auch as plasma, lymph and urine and can be quantified using techniques such as HPLC, mass spectrometry and ELISA. Although these biomarkers are quite stable and eliminate the shortcomings of the very short half-life of

Table 1.3 Key biomarkers of oxidative stress associated with human diseases [55, 57–59]

Disease	Key biomarkers
Alzheimers disease	MDA, HNE, GSH/GSSH ratio, F_2-isoprostanes, NO_2-Tyr, AGE
Atherosclerosis	MDA, HNE, Acrolein, F_2-isoprostanes, NO_2-Tyr
Cancer	MDA, GSH/GSSH ratio NO2-Tyr, 8-OH-dG
Cardiovascular disease	HNE, GSH/GSSH ratio, Acrolein, F_2-isoprostanes, NO_2-Tyr
Diabetes mellitus	MDA, GSH/GSSH ratio, S-glutathionylated proteins, F_2-isoprostanes, NO_2-Tyr, AGE
Ischemia	GSH/GSSH ratio, F_2-isoprostanes
Parkinsons disease	HNE, GSH/GSSH ratio, Carbonylated proteins, Iron level
Rheumatoid arthritis	GSH/GSSH ratio, F_2-isoprostanes
Obesity	MDA, GSH/GSSH ratio, HNE, F_2-isoprostanes

ROS/RNS, the key issue with these biomarkers is that they are either upstream effectors of oxidative stress or more commonly downstream effects [55, 56], and hence cannot be considered an exact reflection of the in vivo redox state.

1.3.2   Approaches to Measure ROS/RNS in Cells

Due to their high reactivity and the cells numerous scavenging mechanisms, in vivo concentrations of ROS/RNS lie in the picomolar to nanomolar range. The detection and quantification of these species in biological systems must therefore be direct, efficient and non-invasive, whilst reporting the redox state in a biological cell. Given the sub-micromolar concentrations of ROS/RNS, another important requirement is sensitivity. The last two decades have witnessed immense focus in the literature on exploring multitude of strategies that would enable real time monitoring of ROS and RNS in biological systems, and the two most extensively used approaches are electron spin resonance (ESR) spectroscopy and fluorescence imaging [60, 61].

One of the earliest analytical approaches developed for the direct detection of ROS/RNS was ESR spectroscopy, which reports on the magnetic properties of unpaired electrons and their molecular environment [62]. The drawbacks associated with the concentration and half-lives of biological free radicals have been efficiently overcome by the use of spin traps. These molecules, mostly nitrone and nitroxide derivatives, bring about the conversion of short lived primary radical species into longer lived radical adducts with characteristic ESR responses [63].

Extensive use of spin traps such as phenyl-t-butylnitrone (**PBN**) and 5,5-dimethylpyrroline-N-oxide (**DMPO**) is reported in the literature, as an aid for the detection of organic radical products of lipid peroxidation and $^\bullet OH$ and $O_2^{\bullet-}$ respectively [61, 62, 64, 65]. More advanced versions of these spin traps, namely 5-diethoxyphosphory-5-methyl-1-pyrroline-N-oxide (DEPMPO) and mito-DEPMPO, have also been developed to eliminate the problems associated with the kinetics and selectivity of PBN and DMPO [63, 66–68]. One of the serious problems associated with these spin traps is that, in presence of cellular reductants like glutathione and ascorbate, the ESR active nitrone-ROS adducts can be rapidly reduced, rendering the adduct ESR inactive and causing underestimation of cellular ROS production. Spin traps such as cyclic hydroxylamines which are less vulnerable to reduction by cellular reductants have been developed and successfully employed in in vivo quantification of ROS [69, 70]. Nevertheless, this strategy has not escaped limitations including instability of spin traps, unwanted reactions with cellular metabolites and lack of spin trap specificity.

1.4 Fluorescence Imaging—an Alternative Approach

Fluorescence imaging is a robust and versatile tool that is applied to visualise structural (such as sub-cellular organelles, protein fibres and membrane-components) and biochemical features (such as pH, redox state, metal ion concentration and temperature) within a biological system [71]. This requires the addition of an exogenous chemical probe that displays the desired fluorescence properties upon binding to the structural component or in response to the analyte of interest [72, 73]. More recently, scientific studies have exploited fluorescence imaging techniques as a strategy to study cellular oxidative capacity [74]. This can be attributed to the specific advantages of fluorescence imaging over other techniques, which include greater sensitivity, high spatial resolution, high specificity and efficiency, precise information on cellular and sub-cellular location, possibility to observe in vivo dynamics, and the synthetic ease of modifying the fluorophore and its photophysics [71].

Owing to the extensive benefits of this technique, the last decade has seen an enormous increase in the development of novel fluorophores particularly suitable for sensing and probing applications. Particular attention has been on the use of these molecules as potential sensors for a variety of cellular analytes including ROS/RNS [74], thiols [75], as well as metals [76].

1.5 Fluorescent Redox Probes

In recent decades, a wide range of fluorescent responsive probes to study redox processes have been developed. Based on their sensing mechanism, these tools can be broadly classified into reaction-based probes and reversible probes. While selective probes usually respond to specific ROS/RNS via a specific irreversible reaction (Fig. 1.4a), reversible probes on the contrary can respond to multiple oxidation-reduction cycles (Fig. 1.4b).

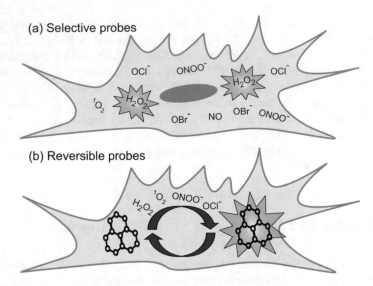

(a) Selective probes

(b) Reversible probes

Fig. 1.4 Two important criteria for the development of redox probes: **a** selective response to a specific ROS/RNS, and **b** reversible response to oxidation-reduction cycles. Reprinted with permission from *Antioxidants and Redox Signalling*, Volume 24, Issue 13, published by Mary Ann Leibert, Inc., New Rochelle, NY

1.5.1   Selective Probes for Detection of ROS/RNS

1.5.1.1   Probes for Hydrogen Peroxide

One of the most extensively utilised fluorescent probes to measure cellular H_2O_2 levels is 2,7 -dichlorodihydrofluorescein (**DCFH**) [77], a non-fluorescent molecule that undergoes a two electron oxidation to the fluorescent 2,7-dichlorofluorescein (Fig. 1.5) [78]. In addition to being oxidised by other ROS/RNS, as well as by enzymatic reactions, recent studies suggest that partial oxidation of **DCFH** can result in a free radical intermediate, which reacts with molecular oxygen, resulting in the formation of superoxide. **DCFH** has further shortcomings such as reduced cellular retention of the oxidised form, auto-oxidation and photosensitivity [79–82]. These problems spurred a quest to identify functional groups or scaffolds that would specifically react with H_2O_2.

The majority of H_2O_2-selective sensors reported to date are based on boronate groups, harnessing the fact that boronic esters are selectively hydrolysed by H_2O_2 [83, 84]. One or more boronic esters are incorporated onto a fluorophore to give a pro-fluorescent probe, which upon hydrolysis by H_2O_2 releases a fluorescent molecule (Fig. 1.6) [85]. Initial attempts in this direction resulted in the preparation of aminocoumarin boronate ester (**AMC**) [84] and Peroxyfluor-1 (**PF1**, Fig. 1.6b) [85], both of which possess a pinacol boronic ester that masks the phenolic oxygen

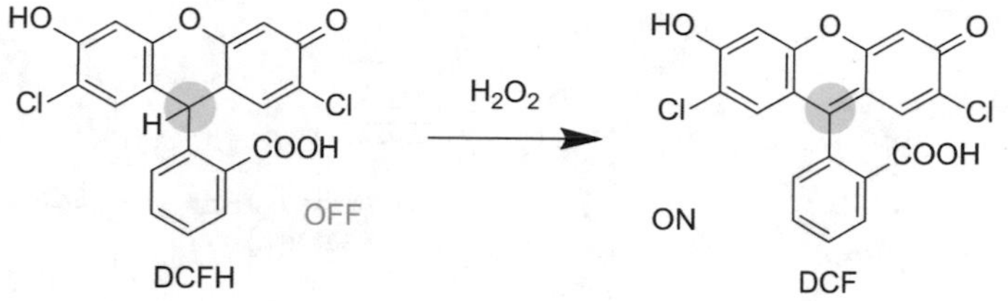

Fig. 1.5 Oxidation reaction of DCFH. Circles indicate the site of redox response. Reprinted with permission from *Antioxidants and Redox Signalling*, Volume 24, Issue 13, published by Mary Ann Leibert, Inc., New Rochelle, NY

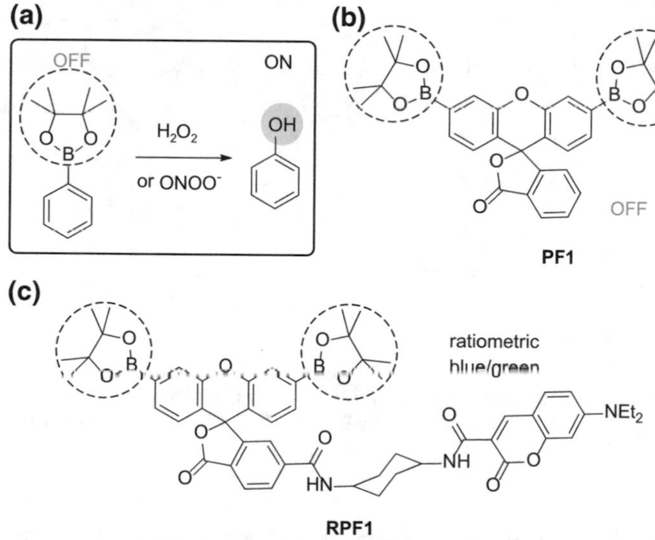

Fig. 1.6 **a** Scheme of boronate cleavage-based detection of H_2O_2 with example probes: **b PF1** [85], **c** ratiometric FRET-based **RPF1** [86]. In all figures, the fluorescence quenching motif is highlighted in a dashed circle, and the moiety enabling fluorescence is shown in a grey circle. Reprinted with permission from *Antioxidants and Redox Signalling*, Volume 24, Issue 13, published by Mary Ann Leibert, Inc., New Rochelle, NY

resulting in low fluorescence. H_2O_2 chemoselectively oxidises the boronic ester to a hydroxyl group, restoring the fluorescence.

Building off this strategy, a myriad of fluorescent probes have been developed, with excitation-emission profiles that span the entire visible spectrum [83, 85, 87–94], including the ratiometric Ratio-Peroxyfluor-1 (**RPF1**, Fig. 1.6c) [86], mitochondrially-localising **MitoPY1** [95], nuclear **NucPE1** [96] and lysosomal **LNB** (Fig. 1.6d) [97]. However, recent work has indicated boronic esters can also be hydrolysed by peroxynitrite ($ONOO^-$), a reaction that proceeds approximately 100-fold faster than with H_2O_2 [98].

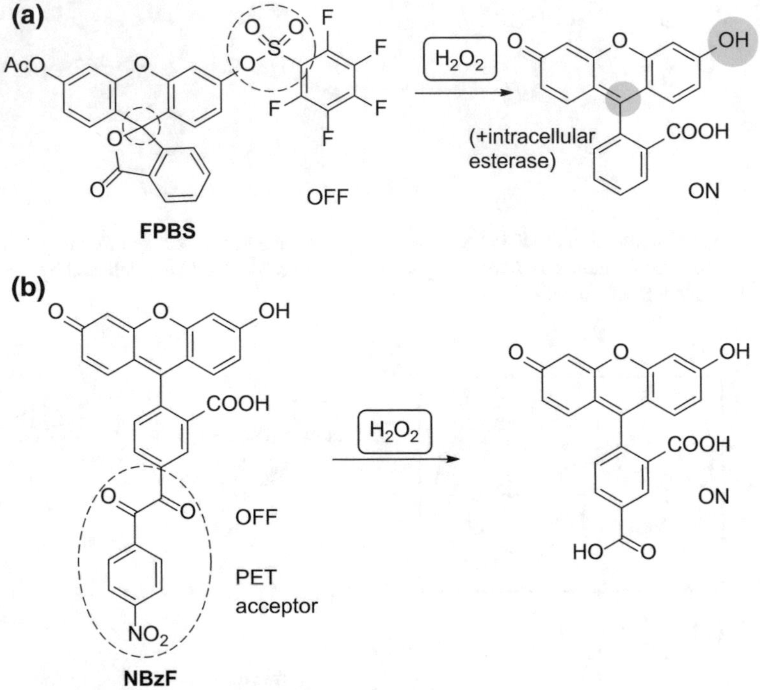

Fig. 1.7 Other irreversible transformations applied for the design of reaction-based H$_2$O$_2$-selective probes: **a** hydrolysis of the sulfonyl ester of **FPBS** [99], **b** H$_2$O$_2$-induced nitrobenzil cleavage in **NBzF** [100]

In addition to the boronic esters, another responsive group exploited for its selectivity towards H$_2$O$_2$-mediated oxidation is the sulfonyl moiety, as in fluorescein-pentafluorobcnzcnesulfonyl ester (**FPBS**, Fig. 1.7a) [99]. Hydrolysis of the sulfonylester by H$_2$O$_2$ results in lactone ring opening. In conjuction with the enzyme-driven loss of acetate, this results in enhanced fluorescence intensity.

A similar approach to H$_2$O$_2$ sensing was demonstrated in fluorescein-benzil (**NBzF**) (Fig. 1.7b) [100]. A Baeyer-Villiger type reaction of the nitro-benzil group with H$_2$O$_2$ yields a benzoic anhydride, which on further hydrolysis releases the fluorescent 5-carboxy fluorescein (Fig. 1.7b). The probe was shown to be selective towards H$_2$O$_2$ over other ROS, but a small fluorescence increase with tBuOOH or ONOO$^-$ was observed [100]. **NBzF** was employed to monitor H$_2$O$_2$ production in phorbol-12-myristate-13-acetate (PMA)-stimulated RAW 264.7 murine macrophages and in A431 epidermoid carcinoma cells stimulated with an epidermal growth factor [100].

1.5.1.2 Probes for Hypochlorous Acid

Most HOCl-selective fluorescent probes developed thus far are based on the HOCl-facilitated rhodamine ring opening mechanism (Fig. 1.8a). This was first demonstrated in **HySOx**, a fluorescent probe selective for HOCl detection [101]. Reaction of HOCl with the thioether moiety in the rhodamine scaffold resulted in ring opening to give the corresponding sulfonate, and a simultaneous increase in fluorescence. **HySOx** was reported to have a good selectivity towards HOCl and was also employed for the detection of HOCl in porcine neutrophils during phagocytosis. Subsequent work yielded thiolactone (**R19S** and **R101S**) and selenolactone (**R19Se**) containing rhodamine probes for HOCl [102].

The first instance of employing HOCl-mediated oxidative hydrolysis of oximes to develop a selective probe was a ratiometric fluorescent probe based on a phenan-throimidazole moiety (**PAI**) bearing oxime functionality (Fig. 1.8b) [103]. Reaction with HOCl resulted in the cleavage of the oxime group to generate the electron withdrawing aldehyde group leading to an intra-molecular charge transfer (ICT) based red-shift in fluorescence. This ratiometric probe was reported to be highly selective for HOCl, and exhibited a 10-fold increase in the ratio (I_{509}/I_{439}) upon treatment with 30 equivalents of HOCl.

Fig. 1.8 **a** Examples of HOCl-induced reactions of ring opening in rhodamine-based probes, exemplified by **HySOx** [101]; **b** response of ratiometric probe, **PAI**, to HOCl [103]. Reprinted with permission from *Antioxidants and Redox Signalling*, Volume 24, Issue 13, published by Mary Ann Leibert, Inc., New Rochelle, NY.eps

1.5.1.3   Probes for Nitric Oxide

The development of fluorescent probes for nitric oxide has taken place in two major domains: organic fluorophores and transition metal-based probes. The first reported organic fluorophore-based nitric oxide probe utilised the PET donor ability of the o-phenylenediamine scaffold, resulting in the fluorescence quenching of a tethered dye. Under aerobic conditions, NO• undergoes rapid oxidation to yield NO+, which reacts with o-phenylenediamine via an N-nitrosation reaction to generate a more fluorescent triazole-containing fluorophore (Fig. 1.9a).

This strategy was employed in the development of diaminofluorescein-2-diacetate (**DAF-2 DA**), a selective probe for NO• bearing diacetyl groups for better cell permeability (Fig. 1.9a), with a more than a 100-fold increase in fluorescence upon reaction with NO• [104]. Although **DAF-2 DA** demonstrated the utility of o-phenylenediamine as an NO responsive scaffold, subsequent probe development that sought to overcome issues related to pH and retention, gave rise to probes with a broad palette of emission colours [107–112].

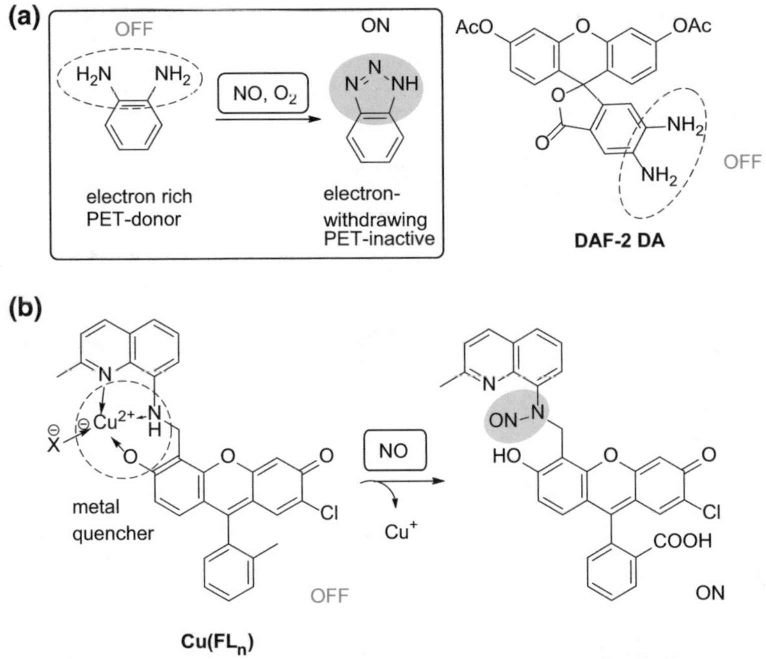

Fig. 1.9 Reaction-based probes for the selective detection of nitric oxide: **a** reaction of NO• with o-phenylenediamine, with example probe **DAF-2 DA** [104]; **b** **Cu(FL$_n$)**, which detects NO• via redox-induced release of the fluorophore [105, 106]. Reprinted with permission from *Antioxidants and Redox Signalling*, Volume 24, Issue 13, published by Mary Ann Leibert, Inc., New Rochelle, NY

Metal complex probes for NO• typically contain a fluorophore coordinated to a ligand bearing a paramagnetic metal ion, which quenches the fluorescence of the fluorophore. Biocompatible Cu(II)-containing probes **Cu(FL$_n$)** have been developed, with ligands bearing quinoline-fluorescein conjugates (Fig. 1.9b) [105, 106]. NO• mediated chemo-selective reduction of copper(II) to copper(I) with simultaneous *N*-nitrosation of fluorescein results in the release of the metal ion from the chelate and in a 16-fold increase in fluorescence.

1.5.1.4 Probes for Peroxynitrite

In addition to the boronate-based probes that have been shown to respond to both ONOO$^-$ and H$_2$O$_2$, probes based on the oxidation of an ethyltrifluoromethylketone group to a dioxirane intermediate have also been developed. **HKGreen-1**, which contains a fluorescein scaffold linked to an ethyltrifluoromethylketone by an aryl-ether bond, exploits this strategy (Fig. 1.10a) [113]. The ketone is oxidised to a dioxirane intermediate that brings about the ether bond cleavage and release of the fluorophore. **HKGreen-2** and **HKGreen-3** probes also respond to ONOO$^-$ via formation of a dioxirane intermediate, unmasking probes fluorescence [114, 115]. Another strat-

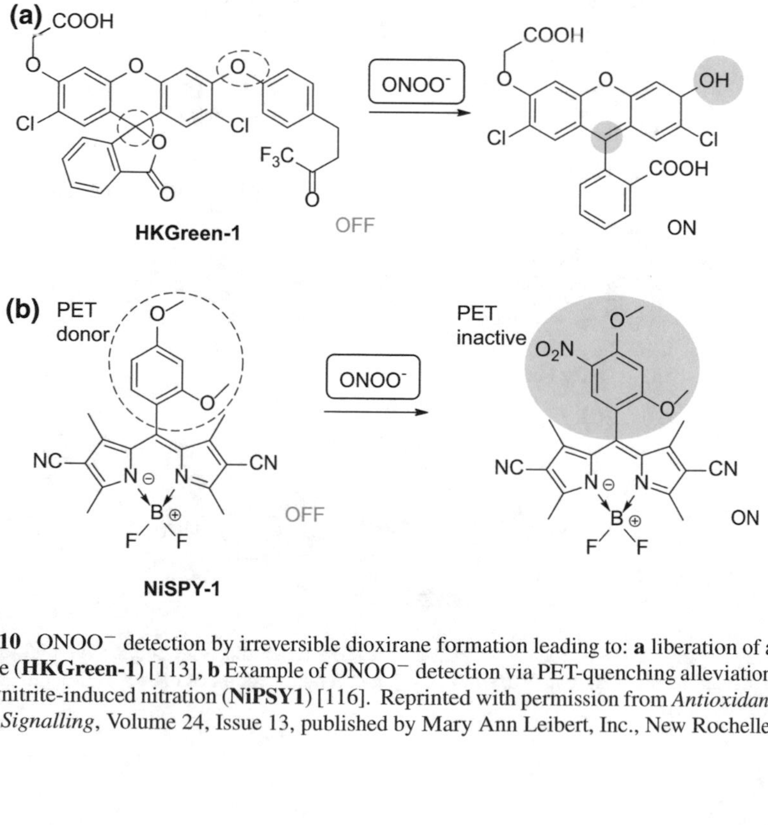

Fig. 1.10 ONOO$^-$ detection by irreversible dioxirane formation leading to: **a** liberation of a fluorophore (**HKGreen-1**) [113], **b** Example of ONOO$^-$ detection via PET-quenching alleviation upon peroxynitrite-induced nitration (**NiPSY1**) [116]. Reprinted with permission from *Antioxidants and Redox Signalling*, Volume 24, Issue 13, published by Mary Ann Leibert, Inc., New Rochelle, NY

egy for ONOO⁻ sensing utilises its potent nitrating ability. Nitration of **NiSPYs** upon reaction with ONOO⁻ removes the PET-quenching, unmasking the probe's fluorescence (Fig. 1.10b) [116].

1.5.1.5 Probes for Superoxide

The superoxide radical anion ($^1O_2^{\bullet-}$) is the primary ROS, a consequence of one-electron reduction of molecular oxygen. Sensing of this ROS particularly focusses on the use of dihydroethidine (**DHE**, Fig. 1.11a) through a mechanism that has been shown by Kalyanaraman et al. to involve the formation of 2-hydroxyethidium (**2-HE**) following a reaction between superoxide and hydroethidine [117, 118]. A different sensing mechanism is concerned with the oxidation of non-fluorescent 2,3-dihydrobenzothiazoles (Fig. 1.11b), for example in **H.Py.Bzt** (2-(2-pyridil)-benzothiazoline) to corresponding strongly fluorescent benzothiazole derivatives [119].

Another strategy involving the deprotection of benzenesulfonate derivatives. for example in **BESSo** (Fig. 1.11c) to sense H_2O_2 has also been applied for the detection

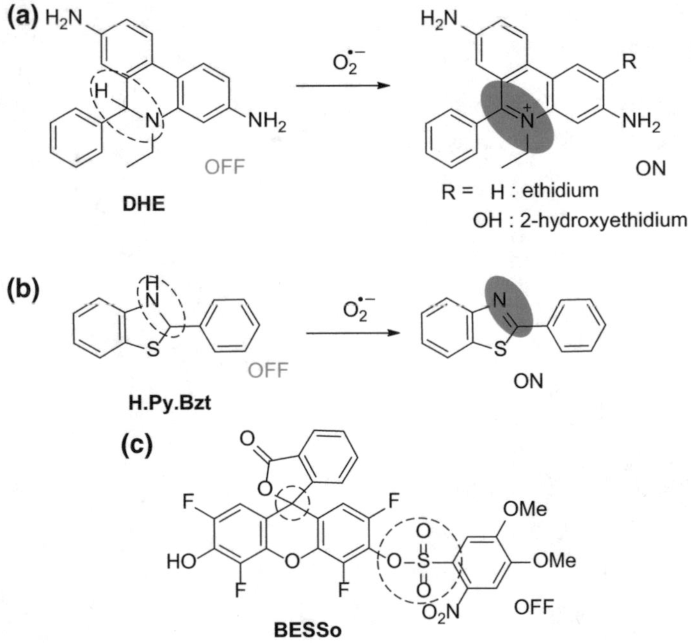

Fig. 1.11 Superoxide detection by oxidation of **a DHE** to ethidine and 2-hydroxyethidium [117, 118] **b** benzothiazole moiety in **H.Py.Bzt** [119], and **d** deprotection of benzenesulfonate group in **BESSo** [120, 121]

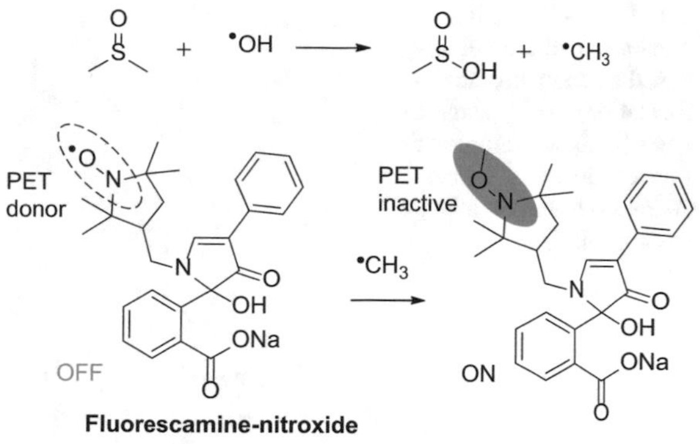

Fig. 1.12 Mechanism of •OH-mediated production of •CH$_3$ which brings about PET-quenching alleviation upon radical oxidation of **Fluorescamine-nitroxide** [124]

of superoxide, this indicates unclear preference of such groups in their reactivity towards ROS [120, 121].

1.5.1.6 Probes for Hydroxyl Radical

Following the work of Blough and colleagues who first initiated the use of nitroxide-conjugated fluorescent sensor [122, 123], Pou and co-workers developed **Fluorescamine-nitroxide** (Fig. 1.12) for sensing •OH in biological systems [124]. These sensors consist of a PET quenching TEMPO moiety attached to a fluorophore. Introduction of •OH in the presence of DMSO resulted in the generation of methyl radicals (•CH$_3$) which then react with the nitroxide and convert it into o-methylhydroxylamine derivative with enhanced fluorescence emission.

Several other nitroxide based probes have been reported, such as **TEMPO-BDP** [125], but significant background fluorescence and other side reactions caused by the presence of DMSO along with high reactivity of nitroxide itself have limited the use of these probes for •OH sensing in biological systems.

1.5.1.7 Probes for Singlet Oxygen

Probes for singlet oxygen have been designed based on the reactivity of the anthracene moiety towards $^1O_2$, to form an endoperoxide (Fig. 1.13a). The Nagano group developed 9,10-diphenylanthracene (DPA) conjugated xanthene probes (**DPAXs**, Fig. 1.13b) for fluorescent detection of $^1O_2$ [126]. $^1O_2$-mediated endoperoxide formation results in significant reduction of the PET quenching by the DPA scaffold and therefore efficient emission from the xanthene fluorophore. Faster singlet oxygen-responsive xanthene-based probes have been developed (**DMAX**) that employ a 9,10-dimethylanthracene trap [127].

Fig. 1.13 **a** Detection
mechanism of singlet oxygen
based on endoperoxide
formation within aromatic
rings [126]. **b** example of
fluorescein-based probes
(**DPAXs**) based on this
strategy [127]

1.5.2   *Reversible Probes for Redox Sensing*

Although the scientific community has turned greatest attention towards the development of selective reaction-based probes, such probes are not able to distinguish
between transient bursts in ROS production typical of physiological events and
chronically-elevated ROS levels, characteristic of pathological oxidative stress. This
is because both situations can result in high ROS measurement at a single timepoint.
In order to make this distinction, monitoring time-resolved changes in the redox state
of the cells is essential. This ability solely depends on the use of reversible probes,
which can cycle back and forth with successive oxidation and reduction events. Using
such a probe, chronic oxidative stress will be distinguished by high ROS-levels over
time, while transient oxidation events will be imaged as high ROS-levels which
subsequently decrease.

Reversible redox sensing abilities have been successfully developed in a set of
redox-responsive fluorescent proteins which possess vital properties such as excellent photostability, bio-compatibility, and ease of intracellular targeting. For example, **HyPer** selectively and reversibly responds to H_2O_2 with a ratiometric emission
change [128]. This and other fluorescent protein-based redox probes have been successfully used in a variety of in vivo models, and continue to deliver information
that was previously beyond the access of scientific community [129]. However, such
probes often require an invasive and usually laborious genetic modification of the
system, and so cannot be applied to a variety of samples, nor will they have potential
clinical applications. In this context, small molecule probes are very promising in
overcoming the intrinsic limitations of genetically-encoded probes. The reversible
redox probes discussed in this section have been classified according to the redox
responsive group.

1.5.2.1 Nitroxide-Based Probes

To enable fluorescent detection, the nitroxide moiety can be covalently tethered to a fluorophore of choice via a suitable linker. The nitroxyl free radical will quench the fluorophores fluorescence [130], which can be restored upon reversible reduction to the diamagnetic hydroxylamine, or irreversible formation of alkoxyamine derivatives (Fig. 1.14a). Since fluorescence is activated upon reduction, these probes are often referred to as profluorescent nitroxides, by analogy to prodrugs. The first proof-of-concept nitroxyl-based fluorophore, **NO-naphthalene** (Fig. 1.14b) [123], exhibits a 10-fold increase in fluorescence quantum yields upon reduction. This strategy has also been applied to other fluorophores in the development of **NO-dansyl** and **NO-perylene** probes [131].

While the nitroxyl radical hydroxylamine redox couple is reversible, the reversibility of nitroxide-based probes is generally not reported. Reversibility of response has been demonstrated for **fluorescein-TEMPO** (Fig. 1.14b), for which one-electron reduction by excess hydrazine hydrate gave rise to an increase in fluorescence, which

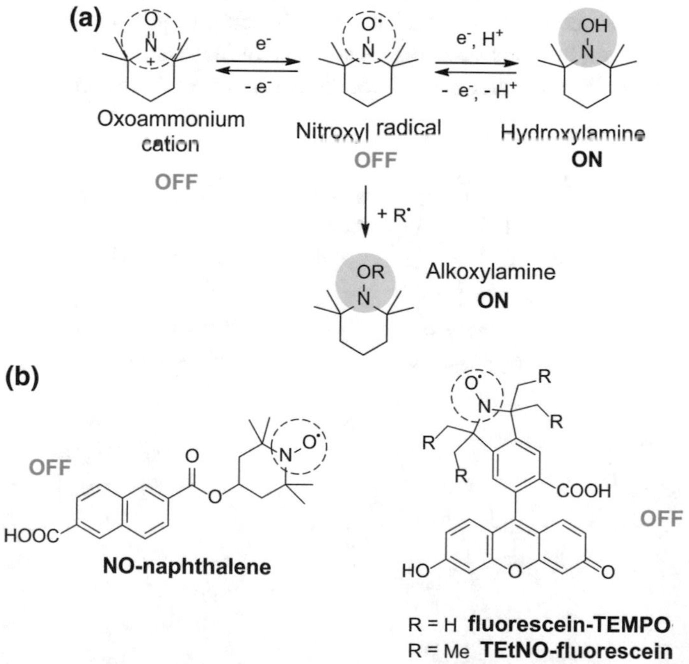

Fig. 1.14 a Possible redox reactions of nitroxyl radicals in biological media. Circles indicate sites of redox response, with fluorescence quenching motifs highlighted in dashed circles and moieties enabling fluorescence shown in grey circles; **b** selected examples of nitroxide-based redox-responsive fluorescent probes, with the fluorescence quenching nitroxyl radical shown in the dashed circle [123, 130, 132–134]. Reprinted with permission from *Angewandte Chemie*, Volume 55, Issue 5, published by John Wiley & sons, Inc.

could then be reversed by air re-oxidation [132, 135]. While this re-oxidation persisted over three redox cycles, with each cycle there was a slight increase in the basal fluorescence intensity, indicative of irreversible destruction of a fraction of the radical.

1.5.2.2 Quinones

The quinone/hydroquinone reversible redox couple (Fig. 1.15a) has been widely studied for almost hundred years [136] and is therefore an obvious choice in the design of redox-responsive optical probes. Quinone-based probes employ a strategy of luminescence quenching by PET from the luminophore in the excited state to the electron-poor quinone motif, as widely reported for porphyrin-based systems [137, 138]. By this strategy, a number of reversible quinone-containing redox sensors based on ruthenium complexes have been reported. For example, the red-luminescent probe **[Ru(bpy)**$_2$**(bpy-Q]**$^{2+}$ (Fig. 1.15b) gave rise to a 4-fold increase in luminescence upon electrochemical reduction in acetonitrile ($E_{red} = -0.2$ V) [139].

Other ruthenium-quinone complexes have been developed [142, 143], and despite electrochemical reversibility in near biologically-relevant potentials (-50 to -300 mV) and significant luminescence response, none of these complexes was examined in more biologically-relevant aqueous conditions, nor was their chemical reversibil-

Fig. 1.15 a reversible redox reactivity of quinone/hydroquinone pair; **b** example of ruthenium-based redox-controlled luminescent switch [139]; **c** mechanism of intracellular response of **Da-Cy** probe to H$_2$O$_2$/thiol redox pair [140]; **d** reduced non-fluorescent form of **TCA** probe [141]. Reprinted with permission from *Angewandte Chemie*, Volume 55, Issue 5, published by John Wiley & sons, Inc.

ity demonstrated. To ensure a redox-dependent fluorescence response in a biological setting, an intracellular redox-active dopamine was directly attached to a cyanine dye (**DA-Cy**, Fig. 1.15c) yielding an on-off-on probe [140]. H_2O_2 oxidation of the 1,2-hydroquinone moiety of dopamine gave a 20-fold fluorescence quenching, and this effect could be counteracted by addition of thiols, in an irreversible process. Disappearance of **DA-Cy** fluorescence upon oxidative stress and subsequent thiol-dependent recovery in HL-7702, HepG2 and RAW 264.7 cells, as well as in rat hippocampal tissue slices, demonstrates the biological utility of **DA-Cy** to study H_2O_2/thiol redox pair. However, the irreversible reduction upon thiol addition excludes the possibility of imaging more than one oxidation/reduction cycle.

1.5.2.3 Chalcogen-Based Fluorescent Redox Probes

A broad class of reversible redox probes involve sulfur, selenium and tellurium in their sensing groups. These can be further divided into those that sense oxidation through formation of a disulfide, diselenide or ditelluride bridge (dichalcogenides, Fig. 1.16a), and those that involve oxidation of the chalcogen itself to the oxide form (Fig. 1.17a).

Dichalcogenides
The sulfide-disulfide oxidation (exemplified in the cysteine to cystine oxidation) is central to countless biological processes and structures. Ratios of thiol-disulfide (whether GSH/GSSG or cysteine/cystine) within cells are therefore widely accepted to be good indicators of cellular oxidative stress [144], ensuring that disulfide is therefore a suitable redox sensing moiety on which the development of fluorescent redox sensors can be based. Likewise, the selenide-diselenide oxidation plays an important role in biology, such as in the catalytic site of glutathione peroxidase [145]. These redox switches have been employed in the development of a limited number of probes discussed below. One of the first reported reversible redox probes based on disulfides, **carbostyril-Tb**, incorporated a carbostyril chromophore separated from a terbium complex by a hexapeptide linker (Fig. 1.16b) [146]. Upon oxidation, the two cysteine residues in the linker form a disulfide bridge that brings the carbostyril and Tb close enough to enable sensitised luminescence, with constant emission intensity at 400 nm enabling ratiometric readout. The reduction potential of this probe has been reported to be -0.243 mV, which lies well within the biological range. While this probe was not tested in biological systems, the authors identify the ease with which the reduction potential can be tuned by modification of the linker. The probe **FSeSeF**, which consists of two fluorescein molecules linked by a diselenide bridge (Fig. 1.16b), utilises an approach similar to the sulfide-disulfide oxidation [75].

Chalcogenoxides
Another strategy for developing fluorescent redox sensors has been to employ chalcogens (S, Se and Te), which can be readily and reversibly oxidised to the respective chalcogenoxides (sulfoxides, selenoxides and telluroxides; Fig. 1.17a). The strategy of employing chalcogen-chalcogenoxide oxidation was pioneered by the Han group

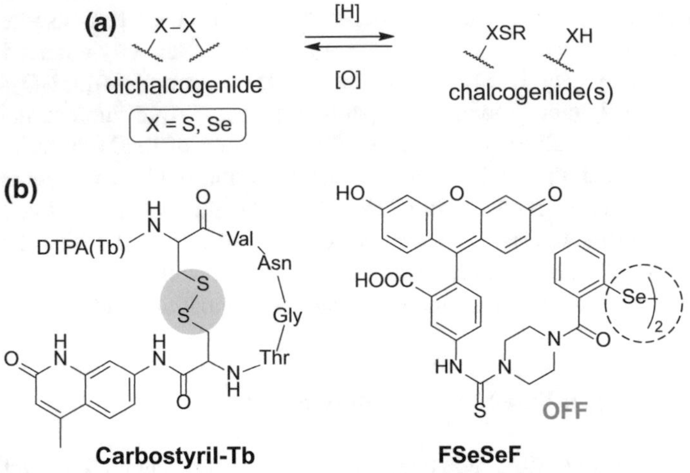

Fig. 1.16 Reversible redox transformations of **a** dichalcogenide and **b** examples of dichalcogenide-type probes [75, 146, 147]. Reprinted with permission from *Angewandte Chemie*, Volume 55, Issue 5, published by John Wiley & sons, Inc.

following the decoding of the catalytic sites of GPx, which showed that selenium in the catalytic pockets reacts with ROS to form selenoxides [145]. The first such probe, **Cy-PSe** (Fig. 1.17b), a near IR emitter, was based on photoinduced electron transfer (PET) between a cyanine signal transducer and a phenyl selenium modulator [148]. Upon oxidation to Se = O, PET quenching is alleviated, resulting in a turn-on in fluorescence. **Cy-PSe** was reported to be selectively oxidised by peroxynitrite and reduced by glutathione and cysteine. The probe was used to measure peroxynitrite in activated mouse macrophages, and its reversibility in biological systems was demonstrated.

A number of further probes have been developed based on this strategy such as **MPhSe-BOD** [151], **diMPhSe-BOD** [151], **HCSe** [152]. and **NI-Se** [153]. This strategy has also been extended towards the development of probes based on other chalcogenides—tellurium (**2Me TeR** [149], Fig. 1.17c) and sulfur ($[\mathbf{Ru(bpy)}_3^{2+}]$-**PTZ** [150], Fig. 1.17d).

1.5.2.4 Nicotinamides and Flavins

Biological systems exhibit complex mechanisms of redox regulation, but the majority of regulatory systems utilise members of the vitamin B group nicotinamides (B3) and flavins (B2). These vitamins, and particularly their nucleotide derivatives, nicotinamide adenine dinucleotide (NAD), flavin adenine dinucelotide (FAD), and flavin mononucleotide (FMN), act as redox active cofactors and coenzymes in cellular redox reactions (Fig. 1.18a, b). The first reported use of such vitamins in redox sen-

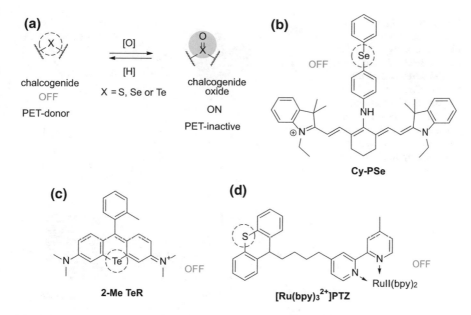

Fig. 1.17 a Reactivity of reversible chalcogenide /chalcogenoxide redox couple and **b, c** selected examples of this type of probes based on **b** Se [148], **c** Te [149] and **d** S [150]. Reprinted with permission from *Angewandte Chemie*, Volume 55, Issue 5, published by John Wiley & sons, Inc.

sors was a PET-based redox molecular switch consisting of a perylene scaffold linked to nicotinamide, **Perylene-NAD** (Fig. 1.18c), which undergoes a 10-fold increase in emission upon oxidation [154]. This molecular switch can be oxidised by *p*-chloranil and reduced by NaBH$_3$CN reversibly for up to 3 cycles, but the process is not selective for these oxidising and reducing agents. Extensive electrochemical and spectro-electrochemical studies give an insight into the reduction potential of this sensor and the plausible electron transfer mechanisms, and highlight the value of these studies for all redox sensors. There are no other reports of nicotinamide-based fluorescent redox probes and it appears that the scientific community has underestimated the potential of the reversible redox properties of nicotinamide.

One of the first reports utilising the reversible redox properties of flavins came from the Aoki group, which developed Zn^{2+}-tetraazacyclododecane complexes bearing lumiflavin and tryptophan [155]. Although these complexes were designed for use as DNA photolyase mimics, the studies emphasised the photochemistry and redox properties of the flavin scaffold. Later, the same group reported **CMFL-BODIPY**, containing carboxymethylflavin (Fig. 1.18d). This probe could be reduced by Na$_2$S$_2$O$_4$ with a 9-fold decrease in fluorescence emission [156]. The reduction potential of this probe was reported to be −240 mV, similar to that of cellular flavins. Although the probes reversibility and selectivity were not reported, studies in HeLa cells demonstrated the potential of the sensing strategy. Again, the potential of this sensing group has been underutilised.

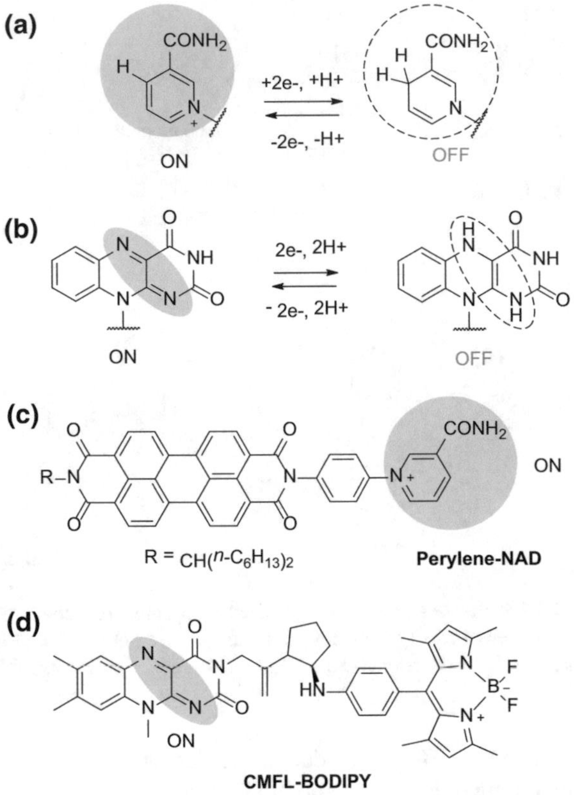

Fig. 1.18 Reversible redox transformations of **a** nicotinamide, **b** flavin moiety and **c,d** selected examples of probes developed on the basis of these redox-responsive groups [154, 155]. Reprinted with permission from *Angewandte Chemie*, Volume 55, Issue 5, published by John Wiley & sons, Inc.

1.6   The Current State of Redox Sensing

This survey of currently available probes reveals the vast potential of the field. In comparison to the field of irreversible probes, many of which have been reported, the development of reversible probes has been relatively slow, and certainly warrants further study.

This literature survey has helped to identify the most promising features of current probes and determine areas that warrant further attention. The nature of the fluorescence change upon oxidation/reduction is crucial in determining the biological utility of a probe. For turn-on (intensity-based) probes, it is important to consider whether the signal is enhanced upon reduction (as for nitroxide-based probes) or upon oxidation (as for most other classes of probes). For the former, probes are likely to be best able to probe questions of hypoxia, or AO efficacy, while the latter will have utility in uncovering new roles of ROS/RNS. However, more promising still are ratiometric probes, for which both oxidised and reduced form can be imaged. In addition to minimising interference from background effects such as probe concentration, ratiometric probes bear the possibility to enable quantification of relative or absolute reduction potential, although this remains to be realised. Despite the promise of probes developed to date, there has been very little work on the rational targeting of probes to specific cell types, or to sub-cellular organelles. Since the redox environment is highly compartmentalised within cells, there is much to be gained from tools that are able to report on location-specific changes in redox state. Another key perspective that is often ignored when designing redox probes is tuning the redox potential of the probes. It is crucial that the probes have a redox potential that falls within the biologically relevant range (-50 to -300 mV). This is essential to ensure that the probe which is introduced within the cells to measure its redox state, doesn't alter the redox potential of the cell itself.

Development of probes utilising the cells' own redox regulators such as nicotinamides and flavins remains underexplored till date. Redox responsive scaffolds such as nicotinamides and flavins not only offer reversible redox properties, but also have reduction potentials well-tuned to the cellular redox events. This suggests that redox probes based on flavins and nicotinamides would not alter the redox potential of the cell. In addition, the reduction potential of flavin and nicotinamide scaffolds can be desirably altered by making suitable structural modifications [157–159]. This ability to tune the reduction potential of a probe will enable development of sets of probes that enable accurate determination of reduction potentials within the cell (analogous to the use of a range of pH indicators to determine acidity). Furthermore, the use of naturally existing molecules eliminates concerns related to cytotoxicity, biological compatibility and cell permeability. Therefore, there is great promise in the development of fluorescent redox probes based on the naturally existing redox molecules.

While fluorescence response is routinely screened in a analysis of a new probe, many studies do not assess the reversibility of the response, the biological sensitivity, or the biological compatibility, but such information is essential: a reversible probe that does not enter cells is not likely to be useful in probing intracellular redox state, nor will a probe with a reduction potential outside the physiologically-relevant range. Key data to gather includes: reversibility over repeated cycles of oxidation-reduction; response to a range of oxidants and reductants (to verify selectivity or global response); stability of signal in the presence of possible interferents such as metal ions, proteins, or pH changes; effect on cell viability; sensitivity to biologically-relevant redox changes as well as the stability, retention and photostability of the probe in time in cellular studies.

Furthermore, the more widespread adoption of fluorescent redox probes will be facilitated by verification that probes can work by more modalities than just confocal microscopy, such as flow cytometry and in plate reader assays. While there is still much work to be done, the promising strategies identified thus far are likely to yield reversible fluorescent sensors of redox state that can be used to distinguish chronic oxidative stress from physiological oxidative bursts not only in cultured cells, but also in vivo studies. The challenge remains to ensure that such probes are put to best use, and that they are employed beyond the laboratory in which they were developed, instead becoming invaluable tools for the redox biology community.

1.7 Objectives

The survey of the literature indicated that there exists a wide array of reaction-based fluorescent redox probes. The aims of the project described in this thesis were to design and synthesise a new class of fluorescent redox-responsive probes. The primary requirement for such tools is reversibility of response to repeated cycles of oxidation and reduction. In order to have utility in biological studies, however, a number of other important aspects must be satisfied, which include:

1. The probe must respond globally to cellular oxidative changes. While reaction-based (irreversible) probes selective for a particular ROS or redox-active pair provide valuable details about the role of redox signalling in a particular biochemical pathway, reversible probes that respond to global oxidative changes are able to report on the overall oxidative capacity of a cell and its recovery from stress.
2. Clear fluorescence response, in which the probe has a high quantum yield and sufficient Stoke's shifts, as well as a large difference in the signal between oxidised and reduced forms. Furthermore, a ratiometric response, in which emission wavelength changes are measured rather than emission intensity, ensures an internal reference that nullifies any concentration, background and instrument-based effects.
3. The fluorescence properties of the probe should be amenable for use with existing imaging technologies, such as confocal microscopy and flow cytometry.
4. Fast reaction kinetics are essential for instantaneous equilibration with the steady state of specific ROS/RNS in the local cellular environment and for maximum spatio-temporal resolution of the signal.
5. Sensitivity of response requires that the fluorescence change is triggered by biologically-relevant redox potentials, or by biologically-meaningful concentrations of ROS/RNS and/or AO.
6. Tunability of redox potential will ensure that probes can be developed that cover the whole range of biologically-relevant redox potentials. Such potentials can differ significantly depending not only on the stage of cellular development, cell type and intracellular localization, but also on the considered biologically-relevant redox pair.
7. Biological compatibility requires the optimization of parameters such as sufficient cellular permeability, and specific sub-cellular localization to enable monitoring of oxidative stress at the organelle level. Furthermore, particularly for reversible probes that offer the potential to monitor cells over time, probes must be non-toxic, and have minimal effect on cellular homoeostasis.
8. The developed probe should be investigated for responsive behaviour in a variety of biological systems in order to understand the extent and limitations of its applicability.

This thesis details the work performed towards the design, synthesis and biological application of reversible redox probes. Chapters 2–4 describe the development of flavin-based redox probes, for imaging both cytoplasmic and mitochondrial oxidative capacity. Chapter 5 presents the strategies investigated for the development of ratiometric redox probes based on nicotinamides. Finally, Chaps. 6–8 discuss the biological applications of the developed probes.

References

1. F.I.F. Benzie, Evolution of antioxidant defence mechanisms. Eur. J. Nutr. **39**, 53–61 (2000)
2. B. Halliwell, Antioxidant defence mechanisms: from the beginning to the end (of the beginning). Free Radic. Res. **31**, 261–272 (1999)
3. S.I. Liochev, Reactive oxygen species and the free radical theory of aging. Free Radic. Biol. Med. **60**, 1–4 (2013)
4. O.I. Aruoma, H. Kaur, B. Halliwell, Oxygen free radicals and human diseases. J. R. Soc. Health **111**, 172–177 (1991)
5. M. Valko, D. Leibfritz, J. Moncol, M.T. Cronin, M. Mazur, J. Telser, Free radicals and antioxidants in normal physiological functions and human disease. Int. J. Biochem. Cell Biol. **39**, 44–84 (2007)
6. R. Gerschman, D.L. Gilbert, S.W. Nye, P. Dwyer, W.O. Fenn, Oxygen poisoning and X-irradiation: a mechanism in common. Science **119**, 623–626 (1954)
7. N.S. Dhalla, R.M. Temsah, T. Netticadan, Role of oxidative stress in cardiovascular diseases. J. Hypertens. **18**, 655–673 (2000)
8. E.H. Sarsour, M.G. Kumar, L. Chaudhuri, A.L. Kalen, P.C. Goswami, Redox control of the cell cycle in health and disease. Antioxid. Redox Signal. **11**, 2985–3011 (2009a)
9. T.C. Jorgenson, W. Zhong, T.D. Oberley, Redox imbalance and biochemical changes in cancer. Cancer Res. **73**, 6118–6123 (2013)
10. H. Kamata, H. Hirata, Redox regulation of cellular signalling. Cell. Signal. **11**, 1–14 (1999)
11. B. D'Autreaux, M.B. Toledano, ROS as signalling molecules: mechanisms that generate specificity in ROS homeostasis. Nat. Rev. Mol. Cell Biol. **8**, 813–824 (2007)
12. C.C. Winterbourn, The challenges of using fluorescent probes to detect and quantify specific reactive oxygen species in living cells. Biochimica et Biophysica Acta (BBA)—General Subjects **1840**, 730–738 (2014)
13. B. Kalyanaraman, V. Darley-Usmar, K.J.A. Davies, P.A. Dennery, H.J. Forman, M.B. Grisham, G.E. Mann, K. Moore, L.J. Roberts, H. Ischiropoulos, Measuring reactive oxygen and nitrogen species with fluorescent probes: challenges and limitations. Free Radic. Biol. Med. **52**, 1–6 (2012)
14. M.D. Brand, C. Affourtit, T.C. Esteves, K. Green, A.J. Lambert, S. Miwa, J.L. Pakay, N. Parker, Mitochondrial superoxide: production, biological effects, and activation of uncoupling proteins. Free Radic. Biol. Med. **37**, 755–767 (2004)
15. T. Finkel, N.J. Holbrook, Oxidants, oxidative stress and the biology of ageing. Nature **408**, 239–247 (2000)
16. B. Halliwell, Generation of hydrogen peroxide, superoxide and hydroxyl radicals during the oxidation of dihydroxyfumaric acid by peroxidase. Biochem. J. **163**, 441–448 (1977)
17. M. Valko, H. Morris, M.T.D. Cronin, *Metals, Toxicity and Oxidative Stress* (2005)
18. C.H. Foyer, H. Lopez-Delgado, J.F. Dat, I.M. Scott, Hydrogen peroxide- and glutathione-associated mechanisms of acclimatory stress tolerance and signalling. Physiol. Plant. **100**, 241–254 (1997)
19. E. Pinto, T.C.S. Sigaud-kutner, M.A.S. Leitão, O.K. Okamoto, D. Morse, P. Colepicolo, Heavy metal-induced oxidative stress in. J. Phycol. **39**, 1008–1018 (2003)
20. J.M. Pullar, M.C.M. Vissers, C.C. Winterbourn, Living with a killer: the effects of hypochlorous acid on mammalian cells. IUBMB Life **50**, 259–266 (2000)

21. L.J. Ignarro, Nitric oxide: a unique endogenous signaling molecule in vascular biology (Nobel Lecture). Angew. Chem. Int. Ed. **38**, 1882–1892 (1999)
22. R.F. Furchgott, Endothelium-derived relaxing factor: discovery, early studies, and identifcation as nitric oxide (nobel lecture). Angew. Chem. Int. Ed. **38**, 1870–1880 (1999)
23. D. Hirst, T. Robson, Nitric oxide physiology and pathology, in *Nitric Oxide*, ed. by H.O. McCarthy, J.A. Coulter), vol. 704, Chap. 1 (Humana Press, 2011), pp. 1–13
24. S. Moncada, R.M. Palmer, E.A. Higgs, Nitric oxide: physiology, pathophysiology, and pharmacology. Pharmacol. Rev. **43**, 109–142 (1991)
25. M.P. Murphy, M.A. Packer, J.L. Scarlett, S.W. Martin, Peroxynitrite: a biologically significant oxidant. Gen. Pharmacol.: Vasc. Syst. **31**, 179–186 (1998)
26. P.Á.L. Pacher, J.S. Beckman, L. Liaudet, Nitric oxide and peroxynitrite in health and disease. Physiol. Rev. **87**, 315–424 (2007)
27. V. Sosa, T. Moline, R. Somoza, R. Paciucci, H. Kondoh, L.L. Me, Oxidative stress and cancer: an overview. Ageing Res. Rev. **12**, 376–390 (2013)
28. P.H.G.M. Willems, R. Rossignol, C.E.J. Dieteren, M.P. Murphy, W.J.H. Koopman, Redox homeostasis and mitochondrial dynamics. Cell Metab. **22**, 207–18 (2015)
29. M. Schieber, N.S. Chandel, ROS function in redox signaling and oxidative stress. Curr. Biol. **24**, R453–R462 (2014)
30. K. Sinha, J. Das, P.B. Pal, P.C. Sil, Oxidative stress: the mitochondria-dependent and mitochondria-independent pathways of apoptosis. Arch. Toxicol. **87**, 1157–1180 (2013)
31. A.A. Starkov, The role of mitochondria in reactive oxygen species metabolism and signaling. Ann. N. Y. Acad. Sci. **1147**, 37–52 (2008)
32. M. Rojkind, J.A. Dominguez-Rosales, N. Nieto, P. Greenwel, Role of hydrogen peroxide and oxidative stress in healing responses. Cell. Mol. Life Sci. **59**, 1872–1891 (2002)
33. M. Noble, J. Smith, J. Power, M. Mayer-Proschel, Redox state as a central modulator of precursor cell function. Ann. N. Y. Acad. Sci. **991**, 251–271 (2003)
34. T. Finkel, Signal transduction by reactive oxygen species. J. Cell Biol. **194**, 7–15 (2011)
35. C.C. Winterbourn, M.B. Hampton, Thiol chemistry and specificity in redox signaling. Free Radic. Biol. Med. **45**, 549–561 (2008)
36. T. Finkel, From sulfenylation to sulfhydration: what a thiolate needs to tolerate. Sci. Signal. **5**, pe10(2012)
37. E.H. Sarsour, M.G. Kumar, L. Chaudhuri, A.L. Kalen, P.C. Goswami, Redox control of the cell cycle in health and disease. Antioxid. Redox Signal. **11**, 2985–3011 (2009)
38. E.R. Stadtman, B.S. Berlett, Reactive oxygen-mediated protein oxidation in aging and disease. Drug Metab. Rev. **30**, 225–243 (1998)
39. J.S. Dawane, V.A. Pandit, Understanding redox homeostasis and its role in cancer. J. Clin. Diagn. Res. **6**, 1796–1802 (2012)
40. R. Zhu, Y. Wang, L. Zhang, Q. Guo, Oxidative stress and liver disease. Hepatol. Res. **42**, 741–749 (2012)
41. H.K. Vincent, A.G. Taylor, Biomarkers and potential mechanisms of obesity-induced oxidant stress in humans. Int. J. Obes. (Lond) **30**, 400–418 (2006)
42. D. Athanasiou, M. Aguila, D. Bevilacqua, S.S. Novoselov, D.A. Parfitt, M.E. Cheetham, The cell stress machinery and retinal degeneration. FEBS Lett. **587**, 2008–2017 (2013)
43. Y.J.H.J. Taverne, A.J.J.C. Bogers, D.J. Duncker, D. Merkus, Reactive oxygen species and the cardiovascular system. Oxid. Med. Cell. Longev. **2013**, 15 (2013)
44. M. Mohsenzadegan, A. Mirshafiey, The immunopathogenic role of reactive oxygen species in Alzheimer disease. Iran J. Allergy Asthma Immunol. **11**, 203–216 (2012)
45. C. Garcia-Ruiz, J.C. Fernandez-Checa, Redox regulation of hepatocyte apoptosis. J. Gastroenterol. Hepatol. **22**(Suppl 1), S38–42 (2007)
46. M. Ristow, K. Zarse, A. Oberbach, N. Klöting, M. Birringer, M. Kiehntopf, M. Stumvoll, C.R. Kahn, M. Blüher, *Antioxidants prevent health-promoting effects of physical exercise in humans* (Proc. Natl. Acad, Sci, 2009)
47. H. Sies, Oxidative stress: oxidants and antioxidants. Exp. Physiol. **82**, 291–295 (1997)

48. A. Skoumalova, J. Hort, Blood markers of oxidative stress in Alzheimer's disease. J. Cell Mol. Med. **16**, 2291–2300 (2012)

49. D. Montero, G. Walther, A. Perez-Martin, E. Roche, A. Vinet, Endothelial dysfunction, inflammation, and oxidative stress in obese children and adolescents: markers and effect of lifestyle intervention. Obes. Rev. **13**, 441–455 (2012)

50. J. Skrha, Effect of caloric restriction on oxidative markers. Adv. Clin. Chem. **47**, 223–247 (2009)

51. H. Yin, New techniques to detect oxidative stress markers: mass spectrometry-based methods to detect isoprostanes as the gold standard for oxidative stress in vivo. BioFactors **34**, 109–124 (2008)

52. H.K. Kim, A.C. Andreazza, The relationship between oxidative stress and post-translational modification of the dopamine transporter in bipolar disorder. Expert Rev. Neurother. **12**, 849–859 (2012)

53. A.M. Pickering, L. Vojtovich, J. Tower, A.D. KJ, Oxidative stress adaptation with acute, chronic, and repeated stress. Free Radic. Biol. Med. **55**, 109–118 (2013)

54. W.A. Pryor, S.S. Godber, Noninvasive measures of oxidative stress status in humans. Free Radic. Biol. Med. **10**, 177–184 (1991)

55. M.A. Smith, C.A. Rottkamp, A. Nunomura, A.K. Raina, G. Perry, Oxidative stress in Alzheimer's disease. Biochimica et Biophysica Acta (BBA)—Mol. Basis Disease **1502**, 139–144 (2000)

56. I. Dalle-Donne, R. Rossi, R. Colombo, D. Giustarini, A. Milzani, Biomarkers of oxidative damage in human disease. Clin. Chem. **52**, 601–623 (2006)

57. H.N. Xu, J. Tchou, L.Z. Li, Redox imaging of human breast cancer core biopsies: a preliminary investigation. Acad. Radiol. **20**, 764–768 (2013)

58. T. Kullisaar, S. Turk, K. Kilk, K. Ausmees, M. Punab, R. Mandar, Increased levels of hydrogen peroxide and nitric oxide in male partners of infertile couples. Andrology **1**, 850–858 (2013)

59. R. Stocker, J.F. Keaney Jr, Role of oxidative modifications in atherosclerosis. Physiol. Rev. **84**, 1381–1478 (2004)

60. D. Ortiz de Orue, Lucana, Redox sensing: novel avenues and paradigms. Antioxid. Redox Signal. **16**, 636–638 (2012)

61. S.I. Dikalov, D.G. Harrison, Methods for detection of mitochondrial and cellular reactive oxygen species. Antioxid. Redox Signal. **20**, 372–382 (2014)

62. E. Finkelstein, G.M. Rosen, E.J. Rauckman, Spin trapping of superoxide and hydroxyl radical: practical aspects. Arch. Biochem. Biophys. **200**, 1–16 (1980)

63. M.M. Tarpey, D.A. Wink, M.B. Grisham, Methods for detection of reactive metabolites of oxygen and nitrogen: in vitro and in vivo considerations. Am. J. Physiol. Regul. Integr. Comp. Physiol. **286**, R431–44 (2004)

64. A. Keszler, B. Kalyanaraman, N. Hogg, Comparative investigation of superoxide trapping by cyclic nitrone spin traps: the use of singular value decomposition and multiple linear regression analysis. Free Radic. Biol. Med. **35**, 1149–1157 (2003)

65. N. Khan, C.M. Wilmot, G.M. Rosen, E. Demidenko, J. Sun, J. Joseph, J. O'Hara, B. Kalyanaraman, H.M. Swartz, Spin traps: in vitro toxicity and stability of radical adducts. Free Radic. Biol. Med. **34**, 1473–1481 (2003)

66. C. Frejaville, H. Karoui, B. Tuccio, F. Le Moigne, M. Culcasi, S. Pietri, R. Lauricella, P. Tordo, 5-(Diethoxyphosphoryl)-5-methyl-1-pyrroline N-oxide: a new efficient phosphorylated nitrone for the in vitro and in vivo spin trapping of oxygen-centered radicals. J. Med. Chem. **38**, 258–265 (1995)

67. M. Hardy, A. Rockenbauer, J. Vasquez-Vivar, C. Felix, M. Lopez, S. Srinivasan, N. Avadhani, P. Tordo, B. Kalyanaraman, Detection, characterization, and decay kinetics of ROS and thiyl adducts of mito-DEPMPO spin trap. Chem. Res. Toxicol. **20**, 1053–1060 (2007)

68. V. Roubaud, S. Sankarapandi, P. Kuppusamy, P. Tordo, J.L. Zweier, Quantitative measurement of superoxide generation using the spin trap 5-(Diethoxyphosphoryl)-5-methyl-1-pyrroline-N-oxide. Anal. Biochem. **247**, 404–411 (1997)

69. S. Dikalov, M. Skatchkov, E. Bassenge, Quantification of peroxynitrite, superoxide, and peroxyl radicals by a new spin trap hydroxylamine 1-hydroxy-2,2,6,6-tetramethyl-4-oxo-piperidine. Biochem. Biophys. Res. Commun. **230**, 54–57 (1997)

70. S. Dikalov, M. Skatchkov, E. Bassenge, Spin trapping of superoxide radicals and peroxynitrite by 1-Hydroxy-3-carboxy-pyrrolidine and 1-Hydroxy-2,2,6,6-tetramethyl-4-oxo-piperidine and the Stability of Corresponding Nitroxyl Radicals Towards Biological Reductants. Biochem. Biophys. Res. Commun. **231**, 701–704 (1997)

71. J. Lakowicz, *Principles of Fluorescence Spectroscopy* (Kluwer Academic/Plenum Publishers, New York, Boston, Dordrecht, London, Moscow, 1999)

72. M.Y. Berezin, S. Achilefu, Fluorescence lifetime measurements and biological imaging. Chem. Rev. **110**, 2641–2684 (2010)

73. N.J. Emptage, Fluorescent imaging in living systems. Curr. Opin. Pharmacol. **1**, 521–525 (2001)

74. A. Gomes, E. Fernandes, J.L. Lima, Fluorescence probes used for detection of reactive oxygen species. J. Biochem. Biophys. Methods **65**, 45–80 (2005)

75. Z. Lou, P. Li, X. Sun, S. Yang, B. Wang, K. Han, A fluorescent probe for rapid detection of thiols and imaging of thiols reducing repair and H2O2 oxidative stress cycles in living cells. Chem. Commun. **49**, 391–393 (2013)

76. G. He, X. Zhao, X. Zhang, H. Fan, S. Wu, H. Li, C. He, C. Duan, A turn-on PET fluorescence sensor for imaging Cu2+ in living cells. New J. Chem. **34**, 1055–1058 (2010)

77. A.S. Keston, R. Brandt, The fluorometric analysis of ultramicro quantities of hydrogen peroxide. Anal. Biochem. **11**, 1–5 (1965)

78. S. Watanabe, In vivo fluorometric measurement of cerebral oxidative stress using 2'-7'-dichlorofluorescein (DCF). Keio J. Med. **47**, 92–98 (1998)

79. N.W. Kooy, J.A. Royall, H. Ischiropoulos, Oxidation of 2',7'-dichlorofluorescin by peroxynitrite. Free Radic. Res. **27**, 245–254 (1997)

80. C. Rota, C.F. Chignell, R.P. Mason, Evidence for free radical formation during the oxidation of 2-7-dichlorofluorescin to the fluorescent dye 2-7-dichlorofluorescein by horseradish peroxidase: Possible implications for oxidative stress measurements. Free Radic. Biol. Med. **27**, 873–881 (1999)

81. P. Wardman, Fluorescent and luminescent probes for measurement of oxidative and nitrosative species in cells and tissues: progress, pitfalls, and prospects. Free Radic. Biol. Med. **43**, 995–1022 (2007)

82. M. Wrona, K. Patel, P. Wardman, Reactivity of 2,7-dichlorodihydrofluorescein and dihydrorhodamine 123 and their oxidized forms toward carbonate, nitrogen dioxide, and hydroxyl radicals. Free Radic. Biol. Med. **38**, 262–270 (2005)

83. J.C. Sanchez, W.C. Trogler, Polymerization of a boronate-functionalized fluorophore by double transesterification: applications to fluorescence detection of hydrogen peroxide vapor. J. Mater. Chem. **18**, 5134–5141 (2008)

84. L.-C. Lo, C.-Y. Chu, Development of highly selective and sensitive probes for hydrogen peroxide. Chem. Commun., 2728–2729 (2003)

85. E.W. Miller, A.E. Albers, A. Pralle, E.Y. Isacoff, C.J. Chang, Boronate-based fluorescent probes for imaging cellular hydrogen peroxide. J. Am. Chem. Soc. **127**, 16652–16659 (2005)

86. A.E. Albers, V.S. Okreglak, C.J. Chang, A FRET-based approach to ratiometric fluorescence detection of hydrogen peroxide. J. Am. Chem. Soc. **128**, 9640–9641 (2006)

87. A.E. Albers, B.C. Dickinson, E.W. Miller, C.J. Chang, A red-emitting naphthofluorescein-based fluorescent probe for selective detection of hydrogen peroxide in living cells. Bioorgan. Med. Chem. Lett. **18**, 5948–5950 (2008)

88. B.C. Dickinson, D. Srikun, C.J. Chang, Mitochondrial-targeted fluorescent probes for reactive oxygen species. Curr. Opin. Chem. Biol. **14**, 50–56 (2010)

89. B.C. Dickinson, C.J. Chang, Chemistry and biology of reactive oxygen species in signaling or stress responses. Nat. Chem. Biol. **7**, 504–511 (2011)

90. L. Du, M. Li, S. Zheng, B. Wang, Rational design of a fluorescent hydrogen peroxide probe based on the umbelliferone fluorophore. Tetrahedron. Lett. **49**, 3045–3048 (2008)

91. F. He, Y. Tang, M. Yu, S. Wang, Y. Li, D. Zhu, Fluorescence-amplifying detection of hydrogen peroxide with cationic conjugated polymers, and its application to glucose sensing. Adv. Funct. Mater. **16**, 91–94 (2006)

92. F. He, F. Feng, S. Wang, Y. Li, D. Zhu, Fluorescence ratiometric assays of hydrogen peroxide and glucose in serum using conjugated polyelectrolytes. J. Mater. Chem. **17**, 3702–3707 (2007)

93. E.W. Miller, C.J. Chang, Fluorescent probes for nitric oxide and hydrogen peroxide in cell signaling. Curr. Opin. Chem. Biol. **11**, 620–625 (2007)

94. D. Srikun, A.E. Albers, C.I. Nam, A.T. Iavarone, C.J. Chang, Organelle-targetable fluorescent probes for imaging hydrogen peroxide in living cells via SNAP-tag protein labeling. J. Am. Chem. Soc. **132**, 4455–4465 (2010)

95. B.C. Dickinson, C.J. Chang, A targetable fluorescent probe for imaging hydrogen peroxide in the mitochondria of living cells. J. Am. Chem. Soc. **130**, 9638–9639 (2008)

96. B. Dickinson, Y. Tang, Z. Chang, C. Chang, A nuclear-localized fluorescent hydrogen peroxide probe for monitoring sirtuin-mediated oxidative stress responses invivo. Chem. Biol. **18**, 943–948 (2011)

97. D. Kim, G. Kim, S.-J. Nam, J. Yin, J. Yoon, Visualization of endogenous and exogenous hydrogen peroxide using a lysosome-targetable fluorescent probe. Sci. Rep. **5** (2015)

98. A. Sikora, J. Zielonka, M. Lopez, J. Joseph, B. Kalyanaraman, Direct oxidation of boronates by peroxynitrite: mechanism and implications in fluorescence imaging of peroxynitrite. Free Radic. Biol. Med. **47**, 1401–1407 (2009)

99. H. Maeda, Y. Fukuyasu, S. Yoshida, M. Fukuda, K. Saeki, H. Matsuno, Y. Yamauchi, K. Yoshida, K. Hirata, K. Miyamoto, Fluorescent probes for hydrogen peroxide based on a non-oxidative mechanism. Angew. Chem. Int. Ed. **43**, 2389–2391 (2004)

100. M. Abo, Y. Urano, K. Hanaoka, T. Terai, T. Komatsu, T. Nagano, Development of a highly sensitive fluorescence probe for hydrogen peroxide. J. Am. Chem. Soc. **133**, 10629–10637 (2011)

101. S. Kenmoku, Y. Urano, H. Kojima, T. Nagano, Development of a highly specific rhodamine-based fluorescence probe for hypochlorous acid and its application to real-time imaging of phagocytosis. J. Am. Chem. Soc. **129**, 7313–7318 (2007)

102. X. Chen, K.-A. Lee, E.-M. Ha, K.M. Lee, Y.Y. Seo, H.K. Choi, H.N. Kim, M.J. Kim, C.-S. Cho, S.Y. Lee, W.-J. Lee, J. Yoon, A specific and sensitive method for detection of hypochlorous acid for the imaging of microbe-induced HOCl production. Chem. Commun. **47**, 4373–4375 (2011)

103. W. Lin, L. Long, B. Chen, W. Tan, A ratiometric fluorescent probe for hypochlorite based on a deoximation reaction. Chemistry: Eur. J. **15**, 2305–2309 (2009)

104. H. Kojima, N. Nakatsubo, K. Kikuchi, S. Kawahara, Y. Kirino, H. Nagoshi, Y. Hirata, T. Nagano, Detection and imaging of nitric oxide with novel fluorescent indicators: diaminofluoresceins. Anal. Chem. **70**, 2446–2453 (1998)

105. M.H. Lim, B.A. Wong, W.H. Pitcock, D. Mokshagundam, M.-H. Baik, S.J. Lippard, Direct nitric oxide detection in aqueous solution by copper(II) fluorescein complexes. J. Am. Chem. Soc. **128**, 14364–14373 (2006)

106. M.H. Lim, D. Xu, S.J. Lippard, Visualization of nitric oxide in living cells by a copper-based fluorescent probe. Nat. Chem. Biol. **2**, 375–380 (2006)

107. Y. Gabe, Y. Urano, K. Kikuchi, H. Kojima, T. Nagano, Highly sensitive fluorescence probes for nitric oxide based on boron dipyrromethene chromophorerational design of potentially useful bioimaging fluorescence probe. J. Am. Chem. Soc. **126**, 3357–3367 (2004)

108. S. Izumi, Y. Urano, K. Hanaoka, T. Terai, T. Nagano, A simple and effective strategy to increase the sensitivity of fluorescence probes in living cells. J. Am. Chem. Soc. **131**, 10189–10200 (2009)

109. H. Kojima, M. Hirotani, N. Nakatsubo, K. Kikuchi, Y. Urano, T. Higuchi, Y. Hirata, T. Nagano, Bioimaging of nitric oxide with fluorescent indicators based on the rhodamine chromophore. Anal. Chem. **73**, 1967–1973 (2001)

110. H. Kojima, Y. Urano, K. Kikuchi, T. Higuchi, Y. Hirata, T. Nagano, Fluorescent indicators for imaging nitric oxide production. Angew. Chem. Int. Ed. Engl. **38**, 3209–3212 (1999)

111. M.J. Plater, I. Greig, M.H. Helfrich, S.H. Ralston, The synthesis and evaluation of o-phenylenediamine derivatives as fluorescent probes for nitric oxide detection. J. Chem. Soc. Perkin Trans. **1**, 2553–2559 (2001)

112. E. Sasaki, H. Kojima, H. Nishimatsu, Y. Urano, K. Kikuchi, Y. Hirata, T. Nagano, Highly sensitive near-infrared fluorescent probes for nitric oxide and their application to isolated organs. J. Am. Chem. Soc. **127**, 3684–3685 (2005)

113. D. Yang, H.-L. Wang, Z.-N. Sun, N.-W. Chung, J.-G. Shen, A highly selective fluorescent probe for the detection and imaging of peroxynitrite in living cells. J. Am. Chem. Soc. **128**, 6004–6005 (2006)

114. T. Peng, D. Yang, HKGreen-3: a rhodol-based fluorescent probe for peroxynitrite. Org. Lett. **12**, 4932–4935 (2010)

115. Z.-N. Sun, H.-L. Wang, F.-Q. Liu, Y. Chen, P.K.H. Tam, D. Yang, BODIPY-based fluorescent probe for peroxynitrite detection and imaging in living cells. Org. Lett. **11**, 1887–1890 (2009)

116. T. Ueno, Y. Urano, H. Kojima, T. Nagano, Mechanism-based molecular design of highly selective fluorescence probes for nitrative stress. J. Am. Chem. Soc. **128**, 10640–10641 (2006)

117. H. Zhao, S. Kalivendi, H. Zhang, J. Joseph, K. Nithipatikom, J. Vasquez-Vivar, B. Kalyanaraman, Superoxide reacts with hydroethidine but forms a fluorescent product that is distinctly different from ethidium: potential implications in intracellular fluorescence detection of superoxide. Free Radic. Biol. Med. **34**, 1359–1368 (2003)

118. C.D. Georgiou, I. Papapostolou, N. Patsoukis, T. Tsegenidis, T. Sideris, An ultrasensitive fluorescent assay for the in vivo quantification of superoxide radical in organisms. Anal. Biochem. **347**, 144–151 (2005)

119. B. Tang, L. Zhang, L.-L. Zhang, Study and application of flow injection spectrofluorimetry with a fluorescent probe of 2-(2-pyridil)-benzothiazoline for superoxide anion radicals. Anal. Biochem. **326**, 176–82 (2004)

120. H. Maeda, K. Yamamoto, Y. Nomura, I. Kohno, L. Hafsi, N. Ueda, S. Yoshida, M. Fukuda, Y. Fukuyasu, Y. Yamauchi, N. Itoh, A design of fluorescent probes for superoxide based on a nonredox mechanism. J. Am. Chem. Soc. **127**, 68–69 (2005)

121. H. Maeda, K. Yamamoto, I. Kohno, L. Hafsi, N. Itoh, S. Nakagawa, N. Kanagawa, K. Suzuki, T. Uno, Design of a practical fluorescent probe for superoxide based on protection-deprotection chemistry of fluoresceins with benzenesulfonyl protecting groups. Chemistry (Weinheim an der Bergstrasse, Germany) **13**, 1946–1954 (2007)

122. D.J. Kieber, N.V. Blough, Determination of carbon-centered radicals in aqueous solution by liquid chromatography with fluorescence detection. Anal. Chem. **62**, 2275–2283 (1990)

123. N.V. Blough, D.J. Simpson, Chemically mediated fluorescence yield switching in nitroxide-fluorophore adducts: optical sensors of radical/redox reactions. J. Am. Chem. Soc. **110**, 1915–1917 (1988)

124. S. Pou, Y.I. Huang, A. Bhan, V.S. Bhadti, R.S. Hosmane, S.Y. Wu, G.L. Cao, G.M. Rosen, A fluorophore-containing nitroxide as a probe to detect superoxide and hydroxyl radical generated by stimulated neutrophils. Anal. Biochem. **212**, 85–90 (1993)

125. P. Li, T. Xie, X. Duan, F. Yu, X. Wang, B. Tang, A new highly selective and sensitive assay for fluorescence imaging of *OH in living cells: effectively avoiding the interference of peroxynitrite. Chemistry (Weinheim an der Bergstrasse, Germany) **16**, 1834–1840 (2010)

126. K. Tanaka, N. Umezawa, K. Kikuchi, Y. Urano, T. Higuchi, Novel fluorescent probes for singlet oxygen. Angew. Chem. Int. Ed. Engl. **38**, 2899–2901 (1999)

127. K. Tanaka, T. Miura, N. Umezawa, Y. Urano, K. Kikuchi, T. Higuchi, T. Nagano, Rational design of fluorescein-based fluorescence probes mechanism-based design of a maximum fluorescence probe for singlet oxygen. J. Am. Chem. Soc. **123**, 2530–2536 (2001)

128. V.V. Belousov, A.F. Fradkov, K.A. Lukyanov, D.B. Staroverov, K.S. Shakhbazov, A.V. Terskikh, S. Lukyanov, Genetically encoded fluorescent indicator for intracellular hydrogen peroxide. Nat. Meth. **3**, 281–286 (2006)

129. A.J. Meyer, T.P. Dick, Fluorescent protein-based redox probes. Antioxid. Redox Signal. **13**, 621–650 (2010)
130. S.A. Green, D.J. Simpson, G. Zhou, P.S. Ho, N.V. Blough, Intramolecular quenching of excited singlet states by stable nitroxyl radicals. J. Am. Chem. Soc. **112**, 7337–7346 (1990)
131. E. Lozinsky, V.V. Martin, T.A. Berezina, A.I. Shames, A.L. Weis, G.I. Likhtenshtein, Dual fluorophore-nitroxide probes for analysis of vitamin C in biological liquids. J. Biochem. Biophys. Methods **38**, 29–42 (1999)
132. B.J. Morrow, D.J. Keddie, N. Gueven, M.F. Lavin, S.E. Bottle, A novel profluorescent nitroxide as a sensitive probe for the cellular redox environment. Free Radic. Biol. Med. **49**, 67–76 (2010)
133. F. Yu, P. Song, P. Li, B. Wang, K. Han, Development of reversible fluorescence probes based on redox oxoammonium cation for hypobromous acid detection in living cells. Chem. Commun. **48**, 7735–7737 (2012)
134. Y. Liu, S. Liu, Y. Wang, TEMPO-based redox-sensitive fluorescent probes and their applications to evaluating intracellular redox status in living cells. Chem. Lett. **38**, 588–589 (2009)
135. H.-Y. Ahn, K.E. Fairfull-Smith, B.J. Morrow, V. Lussini, B. Kim, M.V. Bondar, S.E. Bottle, K.D. Belfield, Two-photon fluorescence microscopy imaging of cellular oxidative stress using profluorescent nitroxides. J. Am. Chem. Soc. **134**, 4721–4730 (2012)
136. L.F. Fieser, The tautomerism of hydroxy quinones. J. Am. Chem. Soc. **50**, 439–465 (1928)
137. D. Gust, T. Moore, Photosynthetic model systems, in *Photoinduced Electron Transfer III*, ed. by J. Mattay, vol. 159, Chap. 3, (Springer, Berlin, 1991), pp. 103–151
138. M.R. Wasielewski, Photoinduced electron transfer in supramolecular systems for artificial photosynthesis. Chem. Rev. **92**, 435–461 (1992)
139. V. Goulle, A. Harriman, J.-M. Lehn, An electro-photoswitch: redox switching of the luminescence of a bipyridine metal complex. J. Chem. Soc. Chem. Commun., 1034–1036 (1993)
140. F. Yu, P. Li, P. Song, B. Wang, J. Zhao, K. Han, Facilitative functionalization of cyanine dye by an on-off-on fluorescent switch for imaging of H2O2 oxidative stress and thiols reducing repair in cells and tissues. Chem. Commun. **48**, 4980–4982 (2012)
141. W. Zhang, X. Wang, P. Li, F. Huang, H. Wang, W. Zhang, B. Tang, Elucidating the relationship between superoxide anion levels and lifespan using an enhanced two-photon fluorescence imaging probe. Chem. Commun. **51**, 9710–9713 (2015)
142. A.C. Benniston, G.M. Chapman, A. Harriman, S.A. Rostron, Reversible luminescence switching in a Ruthenium(II) Bis(2,2:6,2-terpyridine)-Benzoquinone Dyad. Inorg. Chem. **44**, 4029–4036 (2005)
143. Y.-X. Yuan, Y. Chen, Y.-C. Wang, C.-Y. Su, S.-M. Liang, H. Chao, L.-N. Ji, Redox responsive luminescent switch based on a ruthenium(II) complex Ru(bpy)(2)(PAIDH)(2+). Inorg. Chem. Commun. **11**, 1048–1050 (2008)
144. B. Palmieri, V. Sblendorio, Current status of measuring oxidative stress, in *Advanced Protocols in Oxidative Stress II*, ed. by D. Armstrong, vol. 594, Chap. 1 (Humana Press, 2010), pp. 3–17
145. J.T. Rotruck, A.L. Pope, H.E. Ganther, A.B. Swanson, D.G. Hafeman, W.G. Hoekstra, Selenium: biochemical role as a component of glutathione peroxidase. Nutr. Rev. **38**, 280–283 (1980)
146. K. Lee, V. Dzubeck, L. Latshaw, J.P. Schneider, De Novo designed peptidic redox potential probe: linking sensitized emission to disulfide bond formation. J. Am. Chem. Soc. **126**, 13616–13617 (2004)
147. K. Xu, M. Qiang, W. Gao, R. Su, N. Li, Y. Gao, Y. Xie, F. Kong, B. Tang, A near-infrared reversible fluorescent probe for real-time imaging of redox status changes in vivo. Chem. Sci. **4**, 1079–1086 (2013)
148. F. Yu, P. Li, G. Li, G. Zhao, T. Chu, K. Han, A near-IR reversible fluorescent probe modulated by selenium for monitoring peroxynitrite and imaging in living cells. J. Am. Chem. Soc. **133**, 11030–11033 (2011)
149. Y. Koide, M. Kawaguchi, Y. Urano, K. Hanaoka, T. Komatsu, M. Abo, T. Terai, T. Nagano, A reversible near-infrared fluorescence probe for reactive oxygen species based on Te-rhodamine. Chem. Commun. **48**, 3091–3093 (2012)

150. F. Liu, Y. Gao, J. Wang, S. Sun, Reversible and selective luminescent determination of ClO-/H2S redox cycle in vitro and in vivo based on a ruthenium trisbipyridyl probe. Analyst **139**, 3324–3329 (2014)
151. B. Wang, P. Li, F. Yu, J. Chen, Z. Qu, K. Han, A near-infrared reversible and ratiometric fluorescent probe based on Se-BODIPY for the redox cycle mediated by hypobromous acid and hydrogen sulfide in living cells. Chem. Commun. **49**, 5790–5792 (2013)
152. S.-R. Liu, S.-P. Wu, Hypochlorous acid turn-on fluorescent probe based on oxidation of diphenyl selenide. Org. Lett. **15**, 878–881 (2013)
153. Z. Lou, P. Li, Q. Pan, K. Han, A reversible fluorescent probe for detecting hypochloric acid in living cells and animals: utilizing a novel strategy for effectively modulating the fluorescence of selenide and selenoxide. Chem. Commun. **49**, 2445–2447 (2013)
154. P. Yan, M.W. Holman, P. Robustelli, A. Chowdhury, F.I. Ishak, D.M. Adams, Molecular switch based on a biologically important redox reaction. J. Phys. Chem. B **109**, 130–137 (2004)
155. Y. Yamada, S. Aoki, Efficient cycloreversion of cis, syn-thymine photodimer by a Zn2+1,4,7,10-tetraazacyclododecane complex bearing a lumiflavin and tryptophan by chemical reduction and photoreduction of a lumiflavin unit. J. Biol. Inorg. Chem. **11**, 1007–1023 (2006)
156. Y. Yamada, Y. Tomiyama, A. Morita, M. Ikekita, S. Aoki, BODIPY-based fluorescent redox potential sensors that utilize reversible redox properties of flavin. ChemBioChem **9**, 853–856 (2008)
157. J.D. Walsh, A.F. Miller, Flavin reduction potential tuning by substitution and bending. J. Mol. Struct. (Thoechem) **623**, 185–195 (2003)
158. C.O. Schmakel, K.S.V. Santhanam, P.J. Elving, Nicotinamide adenine dinucleotide (NAD+) and related compounds. Electrochemical redox pattern and allied chemical behavior. J. Am. Chem. Soc. **97**, 5083–5092 (1975)
159. P. Bourbon, Q. Peng, G. Ferraudi, C. Stauffacher, O. Wiest, P. Helquist, Synthesis, photophysical, photochemical, and computational studies of coumarin-labeled nicotinamide derivatives. J. Organ. Chem. **77**, 2756–2762 (2012)

Chapter 2
Flavin Based Redox Probes

2.1 Flavin as a Redox Responsive Scaffold

As outlined in Sect. 1.7, the design of a fluorescent probe requires the selection of appropriate fluorophores and a suitable redox responsive group. Careful selection of the redox responsive moiety is crucial as it will define the dynamic range of the probe and hence will determine the biological questions that can be addressed. A few important criteria were considered in making this selection:

1. Reduction and oxidation of the sensing group must trigger a change in its fluorescence properties.
2. The redox behaviour of the molecule and hence its fluorescent properties should ideally be reversible; and
3. The redox potential of the molecule must lie well within the biologically relevant range.

A redox-responsive group that fulfilled these criteria was the flavin moiety. Flavins are redox sensitive molecules naturally present in biological systems, as a part of scaffolds such as flavin adenine dinucleotide (FAD), flavin mononucleotide (FMN), riboflavin (vitamin B_2), and are known to play an important role in orchestrating redox processes within a cell (Fig. 2.1) [1].

It is also well established that the redox responsive range of flavins (-150 to -400 mV) lies within the biological range [2]. Another interesting property of the isoalloxazine core of flavins (Fig. 2.2a) is that it has a planar structure in the oxidised form, and is fluorescent, emitting green light. Upon reduction of the molecule specifically at N-1 and N-5 it acquires a bent structure, rendering it non fluorescent (Fig. 2.2b) [3]. Furthermore, a non-hydrogen atom at the N-10 position of the flavin is essential to stabilise the isoalloxazine form over the alloxazine form (Fig. 2.2a),

---

Parts of the text and figures of this chapter are reprinted from *Chemical Communications*, Issue 50, and Organic and Biomolecular Chemistry Issue 24, with permission from the Royal Society of Chemistry.

© Springer International Publishing AG 2018
A. Kaur, *Fluorescent Tools for Imaging Oxidative Stress in Biology*,
Springer Theses, https://doi.org/10.1007/978-3-319-73405-7_2

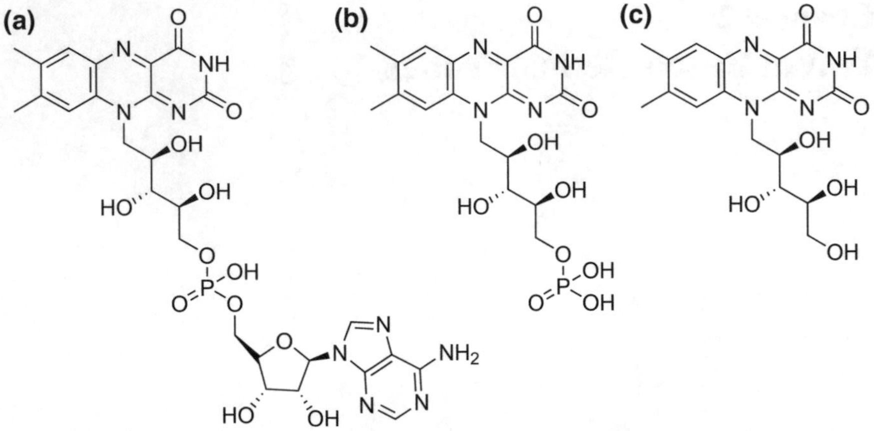

Fig. 2.1 The chemical structures of **a** flavin adenine dinucleotide (FAD), **b** flavin mononucleotide (FMN) and **c** riboflavin

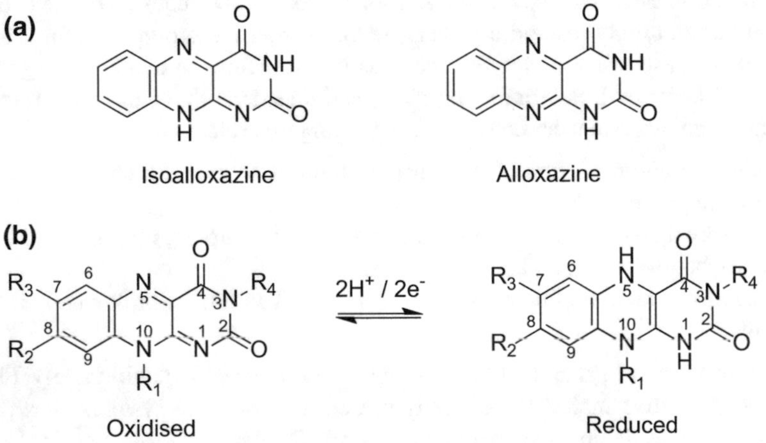

Fig. 2.2 Generic structures of **a** isoalloxazine and alloxazine forms of the flavin scaffold and **b** the oxidised and reduced forms of the isoalloxazine form

and to sustain the reversibility of reduction and oxidation of this molecule [3]. In addition, since flavins are naturally found in cells, their redox chemistry is well-tuned to cellular redox processes, therefore making them ideal for incorporation as a redox responsive group in the probe architecture.

In order to develop a flavin based redox probe, it was decided to suitably modify flavin structure such that it results in a bathochromic shift of its photophysical properties, whilst maintaining its reversible redox behaviour. This would ensure that the probe has fluorescence properties distinct from the naturally occurring flavins [4]. One of the commonly exploited methods of red-shifting the excitation and emis-

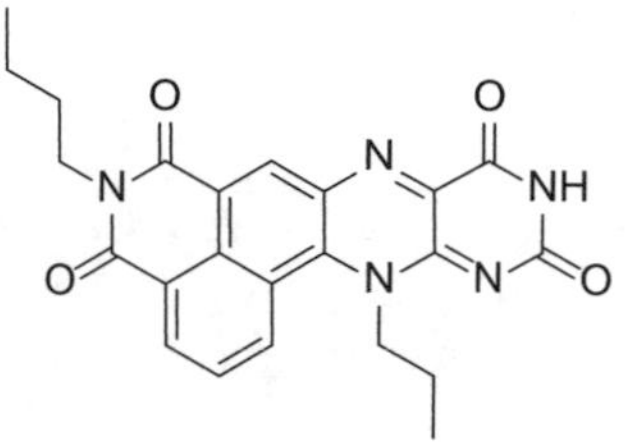

Fig. 2.3 Structure of Naphthalimide flavin based redox sensor 1 (**NpFR1**)

sion wavelengths of a molecule is to increase the overall conjugation [5]. To ensure reversibility of the redox probe it is important to stabilize the isoalloxazine form of the flavin structure over its alloxazine form. This can be achieved by having a non-hydrogen atom at the N-10 position on the flavin scaffold.

2.1.1 Naphthalimide-Flavin Redox Sensor 1—NpFR1

These strategies were applied by a previous Honours student in the group, Jonathan Yeow, who incorporated a naphthalimide fluorophore into the flavin structure. This successfully led to the development of **NpFR1** (napthalimide flavin redox sensor 1, Fig. 2.3).

Preliminary studies (Fig. 2.4) showed that **NpFR1** had absorbance maxima at 395 and 463 nm with molar absorptivity coefficients (log ε) equal to 6.2 and 4 respectively. The emission maximum was found to be at 545 nm. The obtained maxima indicated 27 and 14 nm shifts in absorption and emission respectively compared to riboflavin (Fig. 2.4). Although the bathochromic shift in fluorescence was not as large as expected, **NpFR1** proved to be a good starting point, as its emission spectrum could be distinguished from that of natural flavins by confocal microscopy. Density functional theory (DFT) calculations of basic isoalloxazine ring would be required to further understand the HOMO—LUMO energies and provide useful suggestions about how to further tune the photophysical properties.

Following these preliminary studies, a more detailed understanding of the fluorescence and redox behaviour of **NpFR1** was required before being able to apply the probe for studying cellular redox state using confocal fluorescence microscopy. This chapter details the experiments performed towards rigorous characterisation of the photophysical and redox-responsive properties of **NpFR1**. Furthermore, synthetic strategies aimed towards the development of a mitochondrially-targeted derivative of **NpFR1** are also discussed. Aspects of the work described in this chapter have been published [6, 7].

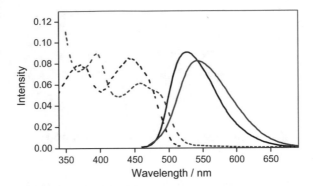

Fig. 2.4 Absorption (dashed line) and fluorescence (solid line) spectra of **NpFR1** (red) and riboflavin (black). Excitation was provided at 405 and 465 nm respectively. Spectra were acquired in HEPES buffer (100 mM, pH 7.4). Reprinted from *Chemical Communications*, Issue 50, with permission from the Royal Society of Chemistry

2.2   Results and Discussion—NpFR1

2.2.1   Redox Responsive Properties

The redox responses of **NpFR1** were assessed by measuring its fluorescence properties when treated with reducing and oxidising agents commonly used in biological experiments, to mimic upper and lower limits of cellular oxidative capacity. Similar to naturally existing flavins, it was possible to reduce **NpFR1** by treating it with mild reducing agents such as sodium dithionite ($Na_2S_2O_4$), sodium cyanoborohydride ($NaBH_3CN$), glutathione (GSH) and dithiothreitol (DTT). As depicted in Fig. 2.5, the oxidised form of **NpFR1** is fluorescent with an emission maximum at 545 nm. Treatments with increasing concentrations of reducing agent $Na_2S_2O_4$, resulted in a decrease in the intensity of this emission maximum and complete reduction of the probe was achieved at a concentration of 100 μM (20 equivalents) of the reducing agent, giving 125-fold lower emission compared to the oxidised form.

In order to assess the re-oxidation of the probe, **NpFR1** was completely reduced by $Na_2S_2O_4$ and then treated with hydrogen peroxide (H_2O_2). The fluorescence properties of the probe indicated that **NpFR1** was oxidised back to its original isoalloxazine form upon treatment with hydrogen peroxide, restoring its absorbance and green fluorescence. Time based analysis of re-oxidation was performed by examining the emission profile of **NpFR1** at 15 s intervals after the treatment of reduced **NpFR1** with H_2O_2. Figure 2.6 indicates that approximately 75% of the original fluorescence was restored within 2 min following the addition of peroxide and about 92% within 4 min (Fig. 2.6). The re-oxidation kinetics of **NpFR1** were found to be similar to those of riboflavin reported in the literature [8, 9]. This means that the probe will be able to detect rapid changes in the oxidative capacity of cells, which is a great

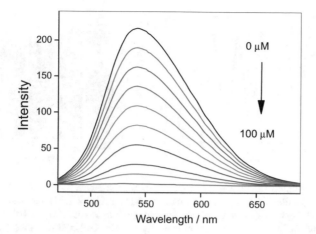

Fig. 2.5 Fluorescence emission of **NpFR1** (5 μM, $\lambda_{ex} = 405$ nm) with the incremental addition of sodium dithionite ($Na_2S_2O_4$). Reprinted from *Chemical Communications*, Issue 50, with permission from the Royal Society of Chemistry

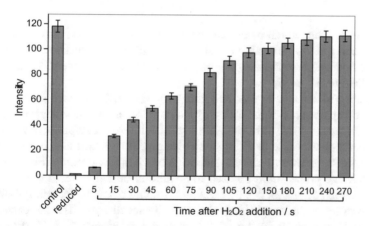

Fig. 2.6 The fluorescence emission from **NpFR1** (5 μM) over time. Bars represent the integrated emission intensity (420 to 650 nm, $\lambda_{ex} = 405$ nm). **NpFR1** was reduced with $Na_2S_2O_4$ (100 μM) and re-oxidised with H_2O_2 (250 μM). All spectra were acquired in HEPES buffer (100 mM, pH 7.4). Error bars represent standard deviation, $n = 3$. Reprinted from *Chemical Communications*, Issue 50, with permission from the Royal Society of Chemistry

advantage over the existing redox probes that require longer treatment durations for complete oxidation [10].

The reversibility of **NpFR1** was assessed by recording the fluorescence emission profile of **NpFR1** upon reduction with $Na_2S_2O_4$, followed by re-oxidation with H_2O_2 and repeating this cycle several times. The reduction—oxidation cycle could be repeated up to 7 times without any significant loss of fluorescent response (Fig. 2.7a). This suggests that **NpFR1** can be used to monitor changes in cellular oxidative

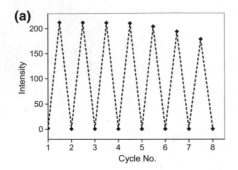
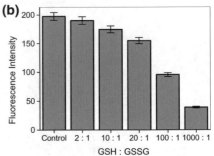

Fig. 2.7 Redox responsive behaviour of **NpFR1**. **a** Fluorescence response of **NpFR1** to cycles of oxidation and reduction. Reduction was achieved with $Na_2S_2O_4$ (50 μM) followed by re-oxidation with 250 μM H_2O_2 **b** Integrated emission intensity (420 to 650 nm, $\lambda_{ex} = 405$ nm) of a 5 μM solution of **NpFR1** in the presence of biologically relevant GSH:GSSG ratios (where the total concentration of GSH + GSSG equals 50 μM in each case) in HEPES buffer (100 mM, pH 7.4). Error bars represent standard deviation, $n = 3$. Reprinted from *Chemical Communications*, Issue 50, with permission from the Royal Society of Chemistry

capacity over time, which would enable a better understanding of redox dynamics within a cell and its physiological and pathological consequences. This probe is therefore a valuable addition to the limited number of reversible redox probes that exist in literature [10–12].

The ratios of reduced to oxidised glutathione (GSH:GSSG) are known to differ between compartments of the cell, with ratios of 2:1 in the endoplasmic reticulum, 10:1 in the nucleus and mitochondria and 100:1 in the cytoplasm [13]. Interestingly, the probe was able to sense small changes when added to solutions containing biologically-relevant reduced and oxidised glutathione ratios (Fig. 2.7b). This shows that this probe could be used to acquire information about the redox state within specific cell organelles by suitably attaching organelle targeting moieties to the probe structure that would bring about specific localisation. This has been applied in the development of mitochondrially-targeted derivative of **NpFR1**, discussed in Sect. 2.2.4.

Since the aim was to develop a probe that responds not only to a single ROS/RNS but to the overall redox environment, re-oxidation of the reduced form of **NpFR1** was assessed in the presence of various biologically relevant ROS/RNS that were produced in situ. For this experiment the probe **NpFR1** (10 μM) was reduced with sodium dithionite (50 μM) and the restoration of its fluorescence emission (indicating re-oxidation) was analysed after treating the reduced probe with 100 μM of various oxidising agents (Fig. 2.8). Re-oxidation was observed to be more rapid with superoxide and hydrogen peroxide, while t-butylperoxide was the slowest. Even in this case, however, more than 95% re-oxidation was achieved within 60 min, indicating that the probe responds to global redox state and not just to individual reactive oxygen species

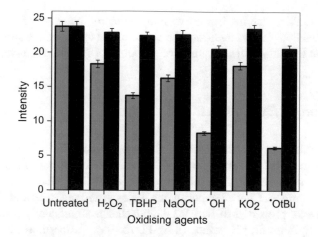

Fig. 2.8 Oxidation of **NpFR1** with various oxidising agents. Bars represent the increase in integrated emission intensity (420 to 650 nm, $\lambda_{ex} = 405$ nm) upon re-oxidation of reduced **NpFR1** (10 μM in 50 μM $Na_2S_2O_4$) immediately (grey) and 60 min after (black) the addition of 100 μM of oxidising agent. Error bars represent standard deviation, $n = 3$. Reprinted from *Chemical Communications*, Issue 50, with permission from the Royal Society of Chemistry

Fig. 2.9 Cyclic voltammogram of **NpFR1** (5 mM concentration) in MeCN with ferrocene as internal standard at 25 °C with a scan rate of 20 mV/s. Voltage given versus SHE. Reprinted from *Chemical Communications*, Issue 50, with permission from the Royal Society of Chemistry

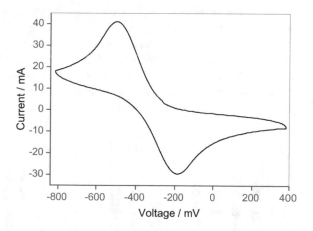

2.2.2 Redox Potential

As outlined in Sect. 1.6, it is very important that the redox potential of a redox probe lies well within the biologically relevant redox potential. Electrochemical studies were performed to assess the redox responsive range of **NpFR1**. Cyclic voltammograms were recorded from degassed 5 mM solution of **NpFR1** in acetonitrile containing tetabutylammonium bromide as an electrolyte and ferrocene as an internal standard.

As calculated from Fig. 2.9, the reduction potential of **NpFR1** is −336 mV versus SHE, and the dynamic range of its redox responsive behaviour extends from −100 mV to −500 mV, which fits in well with the reduction potential of various biological redox reactions [2]. In addition, the shape of the cyclic voltammogram further confirms that the reduction and oxidation reactions of **NpFR1** are reversible.

2.2.3   Control Experiments

In order to assess the applicability of a fluorescent probe in biological systems, it is necessary to assess any effects that pH or biological concentrations of transition metals might have on the fluorescence properties of the probe, or even the cytotoxicity of the probe. Control experiments were carried out to assess these effects. The fluorescence properties of **NpFR1** did not change significantly across the range of biologically relevant pH values (4–9; Fig. 2.10a). Furthermore, as depicted in Fig. 2.10b, the presence of 50 equivalents of common metal ions, much higher than the concentrations normally found in cells, did not alter the emission properties of the probe. These results indicated that other aspects of the biological environment minimally interfere with probe response, therefore reassuring its applicability for biological imaging.

Because of its intended use in cell studies, it was necessary to confirm that **NpFR1** did not have any cytotoxic effects. The cytotoxicity of the probe in 3T3-L1 mouse preadipocytes was assessed using the MTT cytotoxicity assay, which is a colorimetric assay for assessing cell viability (Fig. 2.11). This assay is based on the principle that live cells possess NADPH dependant mitochondrial reductases which are capable of reducing the dye MTT(3-(4,5-dimethylthiazol-2-yl)-2,5-diphenyltetrazolium bromide), a yellow tetrazole into a purple coloured formazan which is then solubilised

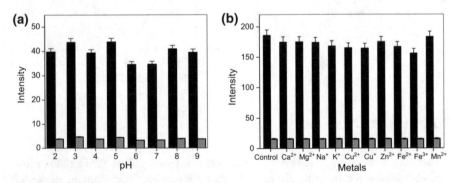

Fig. 2.10 The fluorescence emission from **NpFR1** (5 μM) **a** over a range of pH values and **b** in the presence of common metal ions (100 μM). Bars represent the integrated emission intensity (420 to 650 nm, λ_{ex} = 405 nm) for oxidised (black) and reduced (grey) forms. Error bars represent standard deviation, n = 3. Reprinted from *Chemical Communications*, Issue 50, with permission from the Royal Society of Chemistry

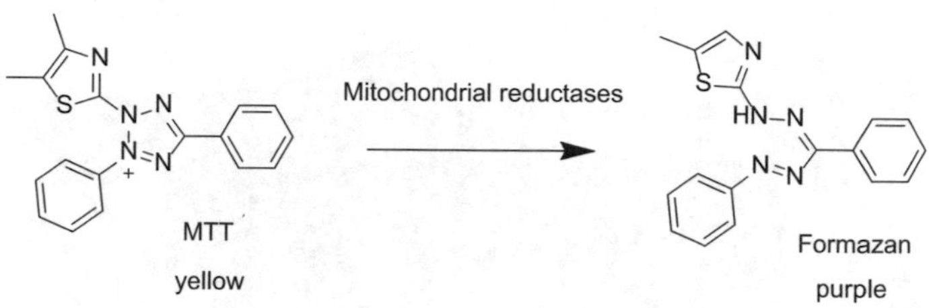

Fig. 2.11 Principle of the MTT cytotoxicity assay

in DMSO and the absorbance is measured at 600 nm. The intensity of absorbance is a measure of cell viablity [14].

The MTT assay of **NpFR1** was performed by growing 3T3-L1 mouse preadipocytes (10,000 cells per well) in a 96 well plate overnight followed by treatment with increasing concentrations of **NpFR1** (ranging from 0–160 μM) for 24 h. The cells were then treated with MTT (20 μL of 2.5 mg/mL solution in PBS) for 4 h after which the media was replaced with 150 μL of DMSO and the absorbance was measured at 600 nm. A plot of the absorbance against the probe concentration gives the half maximal cytotoxic concentration (IC_{50}—probe concentration at which 50% of the cells were viable after 24 h). The IC_{50} value for **NpFR1** was found to be 71 μM.

NpFR1 was subsequently utilised by Matthew Anscomb, an Honours student in the group to measure the differences in oxidative capacity of pre-adipocytes before and after their differentiation into mature adipocytes. **NpFR1** was also employed (50 μM, 2 h) to understand the impact of normal (10 μM) and diabetic glucose (25 μM) concentrations on the oxidative capacity of pre-adipocytes and mature adipocytes [6]. Considering the treatment time and concentrations used in cellular studies it was evident that **NpFR1** did not produce any toxic effects during the biological imaging studies. The ability of **NpFR1** to reversibly respond to changes in the redox environment, and report on oxidative changes in an obesity related model made it an ideal starting point to embark on to developing a mitochondrially targeted redox probe.

2.2.4 Mitochondrially Targeted Redox Probe NpFR2

Microscopy experiments by Matthew Anscomb using **NpFR1** suggested that the probe remained principally in the cytoplasm and responded well to redox changes in the cellular environment. This led to the idea of rationally modifying the probe structure to target it to the mitochondria—the primary source of ROS/RNS [15]. Having similar probes that localise in cytoplasm and mitochondria would enable an understanding of the difference in their oxidative capacity and a delineation between

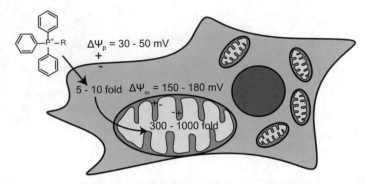

Fig. 2.12 Mechanism for the selective accumulation of lipophilic cation (TPP) across the cell membrane and within the mitochondria

the useful and toxic concentrations of ROS/RNS. In order to target molecules to the mitochondria two essential criteria need to be met-

1. **Positive charge**—The plasma membrane of a cell maintains a low membrane potential ($\Delta\Psi_p = 30-50$ mV; negative inside Fig. 2.12) at all times. As a result, positively charged ions can cross this membrane. Moreover, a substantial potential difference is maintained across the inner mitochondrial membrane ($\Delta\Psi_m = 150-180$ mV; negative inside) [16]. This potential difference is significantly larger than any other area within a cell. As a result positively charged molecules are selectively attracted to the mitochondria resulting in a several hundred-fold accumulation inside the organelle [17]. A widely used strategy for targeting the mitochondria harnesses this remarkable property of the mitochondrial membrane and is based on incorporating cationic moieties within the scaffold.

2. **Lipophilicity**—In addition to a positively charged moiety, sufficient lipophilicity is also vital to achieve selective mitochondrial localisation. A highly localised positive charge would result in ion hydration, thereby rendering the probe cell-impermeable. Scaffolds that possess large hydrophobic surface area, capable of extensive charge delocalisation easily permeate the cell membrane and then accumulate within the mitochondria.

The triphenylphosphonium (TPP) cation has three hydrophobic phenyl residues that delocalise the positive charge present on the central phosphorous atom, and therefore perfectly satisfies the set standards for mitochondrial targeting and has been extensively utilised for this purpose (Fig. 2.12) [17, 18]. Therefore to develop a mitochondrially targeted redox probe, incorporation of a TPP moiety in the **NpFR1** structure was pursued.

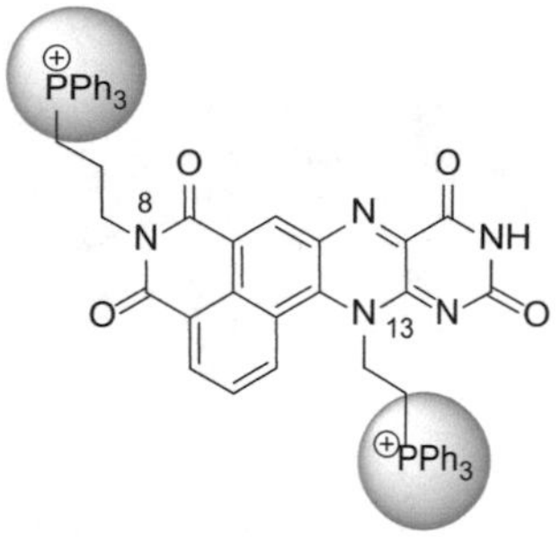

Fig. 2.13 NpFR1 structure
showing the two possible
sites for the incorporation of
a triphenylphosphine moiety

2.2.5 Mitochondrial Targeting Strategies

It was anticipated that modifications of **NpFR1** structure to include a TPP cation
would give mitochondrial localisation, but it is essential to ensure that the photophysical and redox responsive properties are retained. For this purpose, a TPP moiety
needed to be incorporated into the structure *via* an aliphatic linker. As indicated in
Fig. 2.13, two possible sites of incorporation (at $N-8$ and N-13) were identified and
pursued.

2.2.6 Synthesis

2.2.6.1 Targeting *via* N-13

Initially, the attachment of the TPP moiety was attempted at the N-13 tail. One of the
well-tested methods of attaching a TPP moiety is to have a good terminal leaving
group such as a bromide, triflate, tosylate or mesylate that will readily react with
triphenylphosphine to give the corresponding TPP salt. The synthetic scheme was
therefore designed to generate a reactive bromide terminal at N-13. As shown in
Scheme 2.1, alkylation of 4-bromo-1,8-naphthalic anhydride (**1**) with propylamine
was achieved by heating under reflux in ethanol for over 12 h, yielding 4-bromo-
N-propyl-1,8-naphthalimide (**2**). The obtained product was nitrated using sodium
nitrate in sulfuric acid at 0 °C, to afford 3-nitro-4-bromo-N-butyl-1,8-naphthalimide
(**3**) in a procedure analogous to the synthesis of **NpFR1**.

At this step the terminal bromide was introduced by alkylation with
3-bromopropylamine hydrobromide in acetonitrile at room temperature in the presence of a hindered organic base DIPEA, to give the corresponding N-alkylated product (**4**) within 2 h. **4** bore a reactive terminal alkyl bromide, the requisite for TPP
attachment. Reduction of **4** with stannous chloride ($SnCl_2$) in hydrochloric acid (HCl)
resulted in a mixture of products, due to possible polymerisation and intra-molecular
reaction owing to the reactive bromine and amino groups. Considering the difficulty

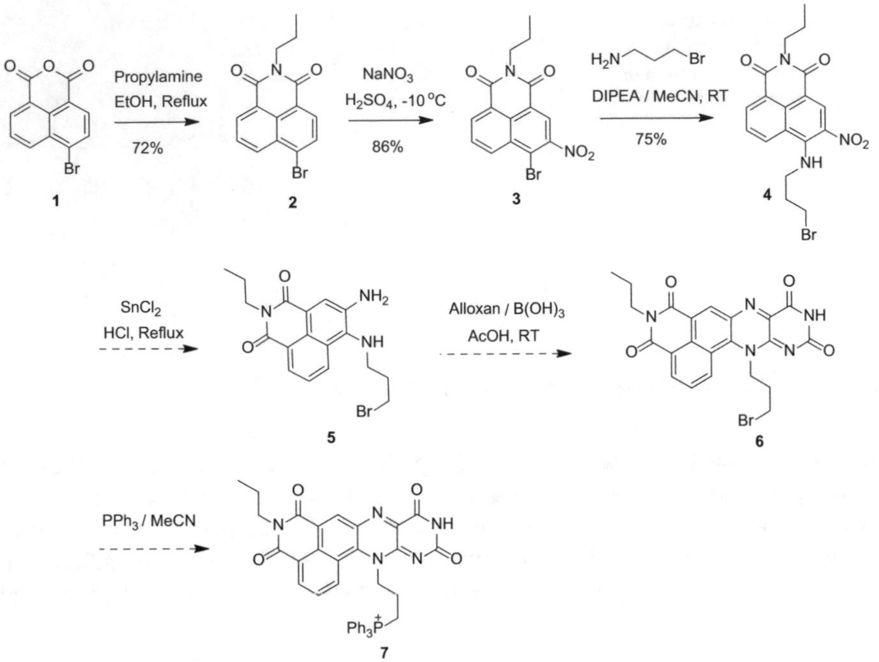

Scheme 2.1 Attempted synthesis of **NpFR2** bearing the mitochondrial targeting group at N-13

of purification and instability of the crude mixture, this site of attachment was not pursued further, and attention was instead focussed on the N-8 position.

2.2.6.2 Targeting *via* N-8

Simultaneous attempts to place the TPP group at the N-8 position proceeded with the alkylation of **1** with 3-bromopropylamine hydrobromide, to introduce the desired terminal bromide (Scheme 2.2). Alkylation was carried out in the presence of DIPEA while heating under ethanol reflux, to give 4-bromo-N-(3-bromopropyl)-1,8-naphthalimide (**8**). Subsequent nitration gave **9**, which was alkylated with propylamine to afford **10**. Considering the synthetic difficulties after reduction of the nitro group that were experienced in the previous attempt, it was decided to pursue the formation of the TPP salt prior to reduction. Room temperature reaction of **10** with triphenylphosphine in acetonitrile did not show any reaction progress. Increasing the reaction temperature and times did not result in satisfactory progress, possibly due to the bulky nature of the TPP group.

Attachment of the TPP at the beginning of the synthetic scheme was therefore attempted. This was initially avoided because of the potential instability of the TPP moiety in the harsh acidic conditions employed later in the synthesis. As seen

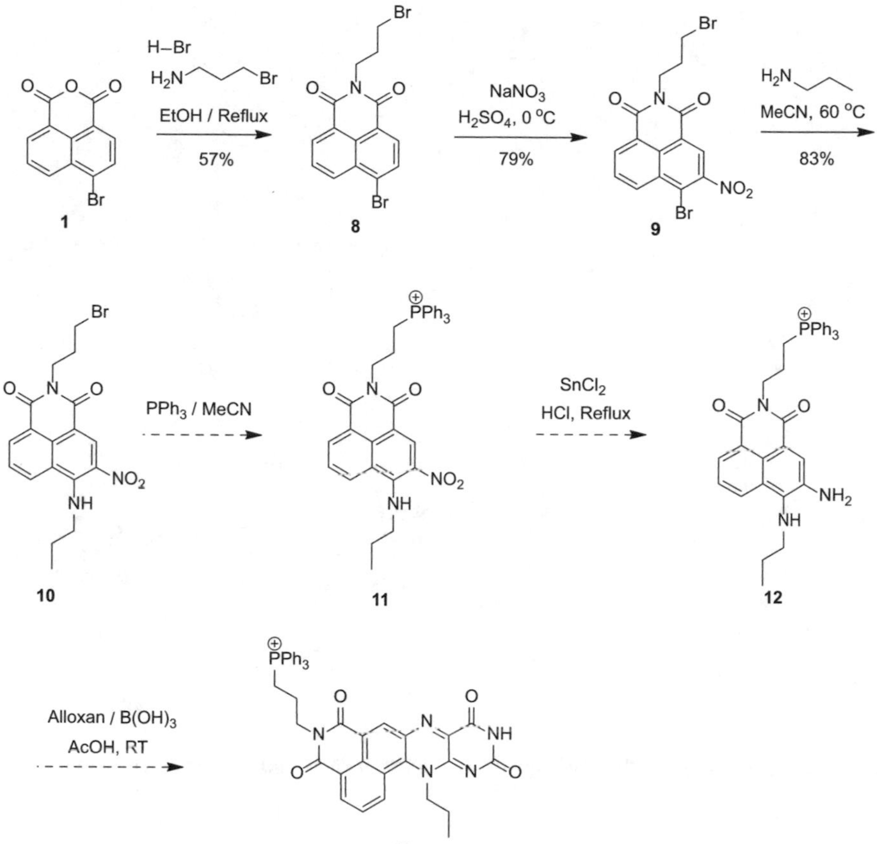

Scheme 2.2   Attempted synthesis of **NpFR2** bearing the mitochondrial targeting group at *N*-8

in Scheme 2.3, the TPP salt of 3-bromopropylamine hydrobromide was prepared by reacting **14** with TPP in acetonitrile and heating under reflux. The resulting (3-aminopropyl) triphenylphosphonium bromide hydrobromide (**15**) was heated under reflux with 4-bromo-1,8-naphthalic anhydride (**1**) in the presence of a hindered base, DIPEA, to yield **16**. Nitration of **16** with sodium nitrate (NaNO$_3$) in sulfuric acid (H$_2$SO$_4$) at −10 °C gave **17**. In contrast to the conditions used in the Sect. 2.2.6.1, nitration was carried out at lower temperature for shorter duration (30 min) to prevent nitration of the phenyl groups of the TPP scaffold. Alkylation of **17** with propylamine using acetonitrile as the solvent at room temperature afforded **18** which was subsequently reduced by stannous chloride (SnCl$_2$) in hydrochloric acid (HCl) to give an *o*-diamino derivative **19**.

Aromatic derivatives that bear adjacent primary and secondary amines have been reported to be suitable substrates for the synthesis of isoalloxazine form of flavins using the alloxan monohydrate method [19]. Due to its instability in air, the formed

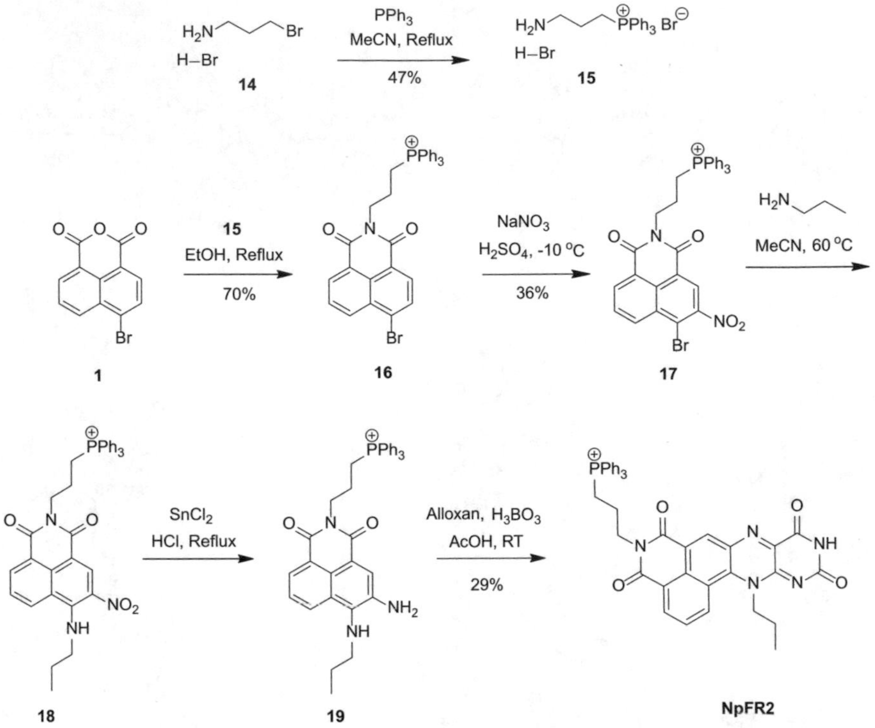

Scheme 2.3 Modified synthesis of **NpFR2** bearing the mitochondrial targeting group at the *N*-8

o-diamino derivative (**19**) was immediately treated with alloxan monohydrate in the presence of boric acid (B(OH$_3$)), using acetic acid as the solvent. Thin layer chromatography (TLC) analysis of the crude reaction mixture indicated the presence of multiple products and was therefore purified using preparative TLC using DCM:MeOH (90:10) as the eluent, to give the desired product **NpFR2** in modest yields.

2.3 Photophysical and Redox Responsive Properties

Having synthesised **NpFR2**, the next step was to characterise its fluorescence and redox sensitive behaviour in comparison to **NpFR1**. The photophysical properties of **NpFR2** were determined in HEPES buffer (100 mM, pH 7.4) to best mimic biological conditions. As expected, the photophysical behaviour of **NpFR2** was comparable to that of **NpFR1**. Figure 2.14 clearly indicates that **NpFR2** has absorption maxima at 470, 489 and 530 nm (Log $\epsilon = 3.8, 5.9$ and 3.1 respectively), and maximum emission

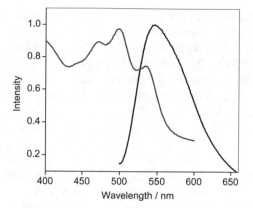

Fig. 2.14 Absorption (red) and fluorescence (black) spectra of **NpFR2**. Excitation was provided at 488 nm. Spectra were acquired in HEPES buffer (100 mM, pH 7.4)

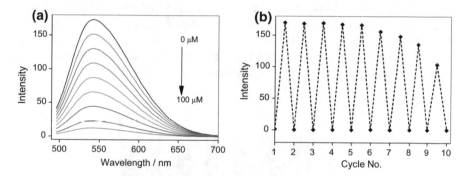

Fig. 2.15 **a** Fluorescence emission spectra of **NpFR2** (10 μM, λ_{ex} = 488 nm) with the incremental addition of sodium dithionite. **b** Fluorescence response of **NpFR2** to cycles of oxidation and reduction. Reduction was achieved with sodium dithionite (100 μM) followed by re-oxidation with 250 μM H_2O_2. Reprinted from Organic and Biomolecular Chemistry, Issue 24, with permission from the Royal Society of Chemistry

at 545 nm. The quantum yield of **NpFR2** fluorescence emission was measured against a fluorescein standard and was calculated to be 0.26.

As in the case of **NpFR1**, the ability of **NpFR2** to respond to the presence of common reducing and oxidising agents was tested. In the oxidised form, the observed fluorescence of **NpFR2** at 545 nm could be attributed to is a planar isoalloxazine ring conformation (Fig. 2.15). As for **NpFR1**, reduction of **NpFR2** could be achieved by treatment with common mild reducing agents such as sodium dithionite, sodium cyanoborohydride, glutathione (GSH) and dithiothreitol (DTT) resulting in a complete loss of fluorescence. As depicted in Fig. 2.15a the oxidised form of **NpFR2** is emissive at 545 nm, while reduction of the probe with 100 μM reducing agent exhibits 115-fold lower emission. Further, the reduced form of **NpFR2** could be re-oxidised by air or by hydrogen peroxide, restoring its planar isoalloxazine form and therefore its original fluorescence emission profile.

Similar to **NpFR1**, the reversibility of **NpFR2** was assessed by analysing the fluorescence profile of **NpFR2** by reducing the probe with sodium dithionite, followed by re-oxidation with H_2O_2 and repeating this cycle several times. One can clearly elucidate from Fig. 2.15b that the reduction-oxidation cycle can be repeated for up to 7 cycles without a significant loss in fluorescence response thus reassuring that the reversibility of the redox responsive properties of **NpFR2** remains unaffected. The fluorescence properties of the probe in conjunction with its redox state and the kinetics of re-oxidation suggest that **NpFR2** follows a similar sensing mechanism to that of riboflavin and **NpFR1**, thus confirming that the TPP tag on the molecule has not affected its fluorescence and redox properties. Figure 2.16 depicts the re-oxidation of reduced **NpFR2** as measured by restoration of its fluorescence, in the presence of various biologically relevant oxidants. Hydrogen peroxide and superoxide were observed to be the most rapid. Nevertheless, **NpFR2** was oxidised almost completely by all oxidants within 30 min. Furthermore, the fact that **NpFR2** could be re-oxidised by a range of diverse oxidants demonstrates that **NpFR2** can be used as a sensor for the global redox state, which can provide complimentary information to sensing a single ROS/RNS. This piece of information is critical for meaningful interpretation of results from biological imaging experiments.

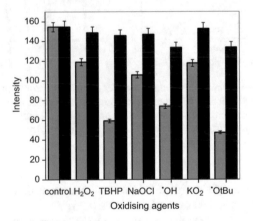

Fig. 2.16 Oxidation of **NpFR2** with various oxidising agents. Bars represent the increase in integrated emission intensity ($\lambda_{ex} = 488$ nm, $\lambda_{em} = 490 - 600$ nm) upon re-oxidation of reduced **NpFR2** (10 μM in 50 μM sodium dithionite) immediately (grey) and 30 min after addition of 100 μM of oxidising agent. Error bars represent standard deviation $n = 3$. Reprinted from Organic and Biomolecular Chemistry, Issue 24, with permission from the Royal Society of Chemistry

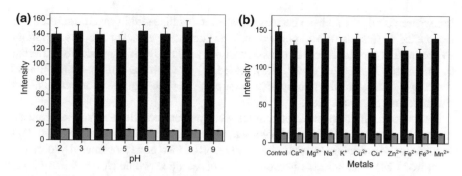

Fig. 2.17 The fluorescence emission of **NpFR2** (10 μM) **a** across a range of pH values and **b** in the presence of common metal ions (100 μM). Bars represent the integrated emission intensity ($\lambda_{ex} = 488$ nm, $\lambda_{em} = 490 - 600$ nm) for oxidised (black) and reduced (grey) forms. Error bars represent standard deviation $n = 3$. Reprinted from Organic and Biomolecular Chemistry, Issue 24, with permission from the Royal Society of Chemistry

2.3.1 Control Experiments

Just as in the case of **NpFR1**, control experiments were performed by assessing the fluorescence properties of the reduced and oxidised forms of **NpFR2** across biologically relevant pH range 4–9 (2.17a) as well as in the presence of biologically available metal ions (2.17b). The emission of neither reduced nor oxidised **NpFR2** was altered significantly by the variations in the pH of the medium or by the presence of common metal ions, therefore confirming that the redox response of the probe is unaffected by other aspects of the biological surroundings.

Furthermore, it is also important to check that addition of the probe to a biological sample does not have any toxic effects on the cells. Standard MTT cell viability assay was carried out by treating HeLa cells with concentrations of **NpFR2** ranging from 0–160 μM for 24 h. The value of IC_{50} was found to be 65 μM for 24 h, which enables the estimation of the concentration of the probe and the duration of treatment that can be applied to biological systems for sensing applications. Therefore, it is crucial to perform biological investigations at concentrations below 65 μM and treatment times shorter than 24 h, to ensure that toxicity would not be a major contributor to the metabolic response.

2.3.2 Mitochondrial Localisation

Having established the redox responsive properties and reversibility of the probe the next step was to test its ability to localise in the mitochondria of cultured cells. Co-localisation experiments are a standard way to determine sub cellular localisation of new fluorescent molecules. These experiments involve co-staining with known tracker dyes that have been established to localise within specific sub cellular organelles, in this case the mitochondria. When choosing a tracker dye, it is important to ensure that the fluorescence emission of the tracker dye is significantly different to that of the fluorescent probe being tested, so that there is minimal interference. An overlay image of the two dyes (tracker and the probe) gives information about the extent of co-localisation.

Mitochondrial co-localisation studies were performed with RAW 264.7 murine macrophages as a representative biological system. Commercially available dyes that localise in the mitochondria (Mitotracker DeepRed FM) and lysosomes (Lysotracker DeepRed FM) were selected because their fluorescence profile ($\lambda_{ex} = 633$ nm, λ_{em} = 650 − 750 nm) was significantly different to that of **NpFR2** ($\lambda_{ex} = 488$ nm, λ_{em} = 490 − 600 nm).

Cells were stained individually with **NpFR2** and Mitotracker DeepRed FM, as well as co-stained with the two dyes, while unstained cells were also maintained as a control (Fig. 2.18). Under excitation by a 488 laser (Fig. 2.18), control cells untreated with the probe showed negligible fluorescence whilst the cells treated with **NpFR2** (10 μM, 15 min) showed significant fluorescence in channel 1 ($\lambda_{ex} = 488$ nm, λ_{em} 490 − 600 nm). Furthermore, **NpFR2**-treated cells exhibited minimal fluorescence in channel 2 ($\lambda_{em} = 650 - 750$ nm), whilst those treated with Mitotracker deep red FM (50 nM, 15 min) or Lysotracker DeepRed (100 nm, 15 min) fluoresce only in channel 2, thus confirming that there is no interference between the fluorescence properties of the probe **NpFR2** and the tracker dyes. Cells were co-stained with **NpFR2** (10 μM, 15 min) and Mitotracker deep red FM (50 nM, 15 min) and the fluorescence images obtained in two distinct channels—channel 1 and channel 2 (Fig. 2.19). The fluorescence images from both the channels were then merged using FIJI (National Institutes of Health), an image processing software. The yellow regions in the merged image suggest that **NpFR2** co-localises with the Mitotracker.

Further confirmation can be obtained by analysing the degree of overlap of fluorescence distribution in two different channels also called Pearson's colocalisation coefficient (PCC) the values of which range from −1 to 1 [20]. PCC value of 1 indicates that the fluorescence intensities of both the images are linearly co-related, giving a linear PCC scatter plot. A PCC value near zero reflects fluorescence distributions that are uncorrelated with one another whereas a value of -1 for indicates the fluorescence intensities are inversely related to each other [20]. The PCC for **NpFR2** was determined using FIJI, to give a value of 0.94 thus confirming that the sub cellular localisation of **NpFR2** is within the mitochondria. Co-staining of RAW 264.7 cells with **NpFR2** and Lysotracker DeepRed FM revealed significantly different localisation regions (Fig. 2.19), accompanied by a poor PCC of 0.10 thus confirming that **NpFR2** does not localise within the lysosomes.

Having confirmed the redox-sensing abilities and mitochondrial localisation of **NpFR2** in cultured cells, the probe was utilised to investigate variations in the mitochondrial redox state of specific haematopoietic cell types isolated from mice. These experiments are discussed in Chap. 7.

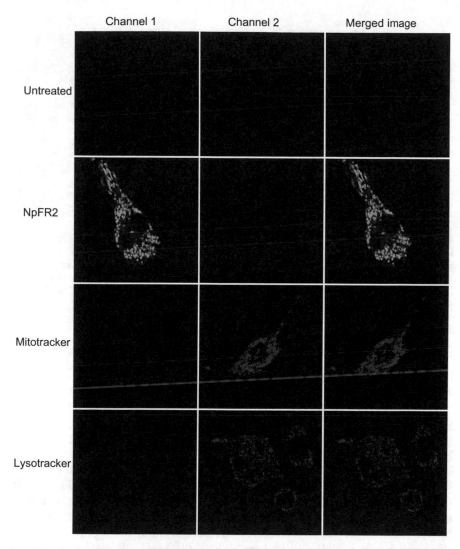

Fig. 2.18 Confocal microscopy images of RAW 264.7 cells untreated and cells treated with **NpFR2** (20 μM, 15 min), Mitotracker DeepRed FM (100 nM, 15 min) and Lysotracker DeepRed (100 nM, 15 min), in channel 1 ($\lambda_{ex} = 488$ nm, $\lambda_{em} = 495 - 600$ nm), channel 2 (100 nM, $\lambda_{ex} = 633$ nm, $\lambda_{em} = 650 - 750$ nm) and merged images of channel 1 and 2. Reprinted from Organic and Biomolecular Chemistry, Issue 24, with permission from the Royal Society of Chemistry

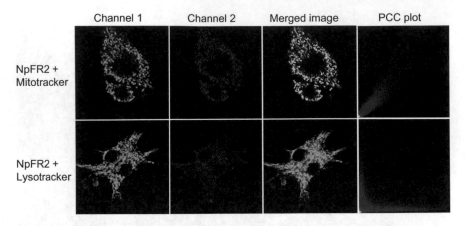

Fig. 2.19 Confocal microscopy images of RAW 264.7 cells co-stained with **NpFR2** (20 μM), and Lysotracker DeepRed (100 nM) for 15 min, in channel 1 ($\lambda_{ex} = 488$ nm, $\lambda_{em} = 495 - 600$ nm), channel 2 (100 nM, $\lambda_{ex} = 633$ nm, $\lambda_{em} = 650 - 750$ nm) and merged images of channel 1 and 2. Intensity correlation plots (PCC) of **NpFR2** (x-axis) versus Mitotracker deep red (y-axis) and Lysotracker deep red (y-axis). Reprinted from Organic and Biomolecular Chemistry, Issue 24, with permission from the Royal Society of Chemistry

2.4 Conclusions

The investigations performed this far establish the use of flavins as sensors of cellular redox state. With the ability to reversibly respond to reduction and oxidation events by changing its fluorescence emission properties and a reduction potential lying within the biological range, flavins exhibit immense potential for use as imaging tools. The photophysical properties of the flavin scaffold can be red-shifted by incorporating another fluorophore within its structure, as seen in case of **NpFR1**. Furthermore, this strategy did not alter the electrochemistry and redox-responsive abilities of the flavin component. The fluorescence and redox behaviour of the probes remained unaffected over a range of pH values and in the presence of biologically-relevant metal ions. Such control experiments are crucial and ensure the applicability of the developed probe in biological systems with minimal or no interference from the cellular environment.

In addition, a mitochondrially-localising derivative **NpFR2** has been developed by introducing a triphenylphosphonium tag *via* an aliphatic linker on the **NpFR1** scaffold. The TPP tag successfully delivered the probe to the mitochondria without altering its fluorescence and redox sensing properties. Having a set of similar probes that localise in different compartments of the cell—the cytoplasm and mitochondria would be valuable to distinguish between the variations in oxidative capacity of mitochondria and cytoplasm. **NpFR1** and **NpFR2** have been further applied in a variety of biological systems to interrogate variations in oxidative capacity within these systems. These experiments have been discussed in Chaps. 6–8.

References

1. V. Massey, Activation of molecular oxygen by flavins and flavoproteins. J. Biol. Chem. **269**, 22459–22462 (1994)
2. J.D. Walsh, A.F. Miller, Flavin reduction potential tuning by substitution and bending. J. Mol. Struct. (Thoechem) **623**, 185–195 (2003)
3. A.J.W.G. Visser, S. Ghisla, V. Massey, F. MÜLler, C. Veeger, Fluorescence properties of reduced flavins and flavoproteins. Eur. J. Biochem. **101**, 13–21 (1979)
4. K. Koenig, H. Schneckenburger, Laser-induced autofluorescence for medical diagnosis. J. Fluoresc. **4**, 17–40 (1994)
5. J. Lakowicz, *Principles of Fluorescence Spectroscopy* (Kluwer Academic/Plenum Publishers, New York, Boston, Dordrecht, London, Moscow, 1999)
6. J. Yeow, A. Kaur, M.D. Anscomb, E.J. New, A novel flavin derivative reveals the impact of glucose on oxidative stress in adipocytes. Chem. Commun. **50**, 8181–8184 (2014)
7. A. Kaur, K.W.L. Brigden, T.F. Cashman, S.T. Fraser, E.J. New, Mitochondrially targeted redox probe reveals the variations in oxidative capacity of the haematopoietic cells. Organ. Biomol. Chem. **13**, 6686–6689 (2015)
8. D.E. Edmondson, T.P. Singer, Oxidation-reduction properties of the 8α-substituted flavins. J. Biol. Chem. **248**, 8144–8149 (1973)
9. V. Favaudon, Oxidation kinetics of 1, 5-dihydroflavin by oxygen in non-aqueous solvent. Eur. J. Biochem. **78**, 293–307 (1977)
10. Y. Yamada, Y. Tomiyama, A. Morita, M. Ikekita, S. Aoki, BODIPY-based fluorescent redox potential sensors that utilize reversible redox properties of flavin. ChemBioChem **9**, 853–856 (2008)
11. R.M. Kierat, B.M. Thaler, R. Kramer, A fluorescent redox sensor with tuneable oxidation potential. Bioorg. Med. Chem. Lett. **20**, 1457–1459 (2010)
12. F. W. Miller, S.X. Bian, C.J. Chang, A fluorescent sensor for imaging reversible redox cycles in living cells. J. Am. Chem. Soc. **129**, 3458–3459 (2007)
13. Y.M. Go, D.P. Jones, Redox compartmentalization in eukaryotic cells. Biochim. Biophys. Acta **1780**, 1273–1290 (2008)
14. J. van Meerloo, G.J.L. Kaspers, J. Cloos, Cell sensitivity assays: the MTT assay, in *Methods in Molecular Biology* , vol. 731, (Clifton, N.J., 2011), pp. 237–245
15. A.A. Starkov, The role of mitochondria in reactive oxygen species metabolism and signaling. Ann. N. Y. Acad. Sci. **1147**, 37–52 (2008)
16. M.F. Ross, T.A. Prime, I. Abakumova, A.M. James, C.M. Porteous, R.A.J. Smith, M.P. Murphy, Rapid and extensive uptake and activation of hydrophobic triphenylphosphonium cations within cells. Biochem. J. **411**, 633–645 (2008)
17. M.P. Murphy, Targeting lipophilic cations to mitochondria. Biochimica et Biophysica Acta—Bioenergetics **1777**, 1028–1031 (2008)
18. A.M. James, H.M. Cochemé, M.P. Murphy, Mitochondria-targeted redox probes as tools in the study of oxidative damage and ageing. Mech. Ageing Dev. **126**, 982–986 (2005)
19. R. Kuhn, K. Reinemund, Über die Synthese des 6.7.9-Trimethyl-flavins (Lumi-lactoflavins). *Berichte der deutschen chemischen Gesellschaft (A and B Series)* **67**, 1932–1936 (1934)
20. K.W. Dunn, M.M. Kamocka, J.H. McDonald, A practical guide to evaluating colocalization in biological microscopy. Am. J. Physiol. Cell Physiol. **300**, C723–C742 (2011)

Chapter 3
FRET Based Ratiometric Redox Probes

To date, only a limited number of reversible redox probes have been reported, most of which are intensity-based, for example **NpFR1** and **NpFR2** reported in the Chap. 2 [1, 2]. In such probes, the fluorescence emission intensity correlates to the oxidative capacity of the environment (Fig. 3.1a). Intensity-based probes are simple and offer an easily interpretable readout. Nevertheless, one needs to consider the fact that changes in the fluorescence intensity of such probes may also arise from variations in probe concentration, probe environment (pH, transition metals) and instrumental factors (such as excitation intensity, emission collection efficiency), which may be falsely interpreted as a reflection of redox changes [3]. Furthermore, it is difficult to conclude whether a lack of fluorescence signal corresponds to the absence of the analyte or the probe itself. New generation ratiometric probes offer an excellent way of overcoming these shortcomings. This chapter details the design and synthetic strategies employed towards the development of a ratiometric redox probe, which is then employed in biological imaging experiments. Aspects of the work described in this chapter have been published in *Chemical Communications* [4].

3.1 Ratiometric Probes

Ratiometric probes report an event through modulation in the response of two different excitation and/or emission maxima. The ratio of these two peaks is calculated rather than the intensity of a single peak (Fig. 3.1b). Ratiometric probes thus offer an internal reference, eliminating accumulation and instrument-based false positives, and therefore enable extraction of quantitative information about the event of interest

Parts of the text and figures of this chapter are reprinted from *Chemical Communications*, Issue 52, with permission from the Royal Society of Chemistry.

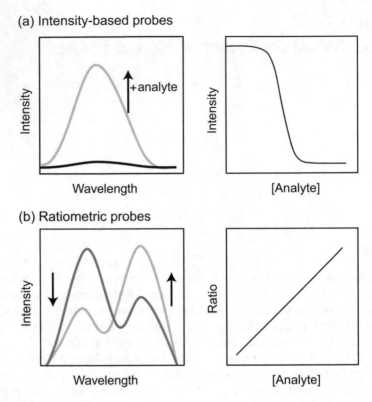

Fig. 3.1 Typical fluorescence response of **a** intensity-based and **b** ratiometric probes

[3]. There is therefore great value in seeking robust methods to develop ratiometric sensors.

3.1.1   Förster Resonance Energy Transfer-FRET

There are a number of different mechanisms enabling a ratiometric fluorescence response, such as intra-molecular charge transfer (ICT) [5, 6], excited state intra-molecular proton transfer (ESIPT) [7] and aggregation induced emission (AIE) [8]. One well established method is Förster Resonance Energy Transfer (FRET)—a distance dependent energy transfer mechanism that operates between two fluorophores for which the emission profile of one fluorophore (referred to as the donor) shows a significant overlap with the excitation profile of the other fluorophore (called the acceptor, Fig. 3.2) [9]. As a consequence of this overlap, the excited donor fluorophore non-radiatively transfers its energy to the acceptor fluorophore, which then fluoresces (Fig. 3.2) [9].

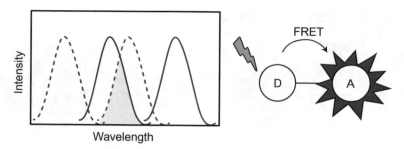

Fig. 3.2 The spectral overlap (shaded yellow) between the donor-emission (blue, solid) and acceptor-excitation (red, dashed) profiles (left) is crucial for the FRET-based energy transfer from an excited donor (D) to the acceptor (A) fluorophore (right)

FRET is a valuable design strategy that can be employed to make a ratiometric responsive probe by incorporating a responsive fluorophore into the structure. Depending on whether the donor or the acceptor is chosen as the responsive group, the fluorescence response can be :

- **Excitation ratiometric**—such probes are excited alternately by two independent excitation wavelengths and the emission at a single wavelength detected (Fig. 3.3a). The ratio of emission intensities at two different excitation wavelengths is used to quantify the signalling event.
- **Emission ratiometric**—these probes are excited with light of a single wavelength and the emission is detected at two independent wavelengths. The ratio of these two emissions can then be successfully applied to assess a particular event (Fig. 3.3b). Simultaneous real time data collection and simplicity of instrumentation make emission ratiometric probes preferable over excitation ratiometric ones.

The efficiency of FRET depends primarily on two factors: the extent of overlap between donor emission and acceptor excitation profiles; and secondly, the distance between the two fluorophores [10]. Therefore, in order to successfully design a FRET-based ratiometric probe, both the fluorophores as well the linker between them must be carefully selected. The aim of this section of the work was to investigate FRET as a strategy to develop a ratiometric redox probe.

3.2 Designing a FRET-Based Redox Probe

The design of a FRET-based ratiometric probe to respond to redox state required judicious selection of three essential components—a redox responsive moiety, a donor-acceptor FRET pair and an appropriate linker, which are discussed below.

(a) Excitation ratiometric FRET-pair

(b) Emission ratiometric FRET-pair

Fig. 3.3 The FRET-sensing mechanism in **a** excitation ratiometric and **b** emission ratiometric sensors

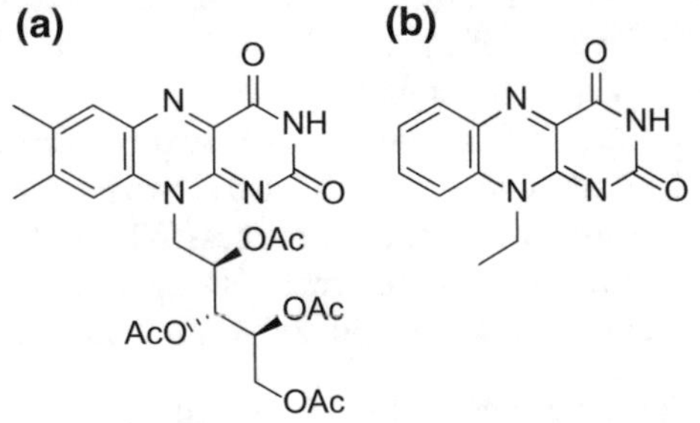

Fig. 3.4 Chemical structures of two flavin derivatives **a** tetraacetylriboflavin and **b** N-ethylflavin investigated for the development of FRET-based redox probe

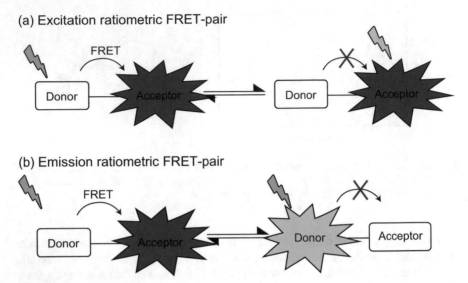

3.2.1   Redox Responsive Moiety

As highlighted in Sect. 2.1, flavins are an ideal choice for the redox responsive group because they play important roles in regulating sub-cellular redox processes and are therefore well-tuned to the biologically relevant redox potential. The fluorescence properties of flavins are indicative of their redox state, with negligible absorbance and emission in the reduced form and green fluorescence upon oxidation. Therefore, to develop a FRET-based redox probe two flavin derivatives were investigated—a naturally occurring flavin derivative (tetraacetylriboflavin, Fig. 3.4a) and a synthetic flavin (N-ethylflavin, Fig. 3.4b).

3.2.2 FRET Pair

Designing a redox probe that is emission ratiometric demanded that the redox responsive flavin moiety be used as the acceptor, and a suitable donor fluorophore be selected. Analysing the absorbance and emission spectrum of flavin, it was understood that a donor fluorophore that would form an ideal FRET pair with flavin should ideally have an emission spectrum ranging from 350 to 480 nm, i.e. the blue region of the spectrum (Fig. 3.5). There is a broad array of blue fluorescent molecules in the literature [11], amongst which coumarins have been extensively studied and their photophysical properties well established [12]. Minor synthetic modifications to the coumarin scaffold can help tune the photophysical properties of the molecule. Figure 3.5 demonstrates the significant overlap of the emission spectrum of 7-diethylaminocoumarin with the excitation spectrum of a flavin, confirming that coumarin would be an ideal choice as a donor molecule to form a FRET pair with flavin.

Therefore, three coumarin derivatives—7-amino-4-methylcoumarin, coumarin-3-carboxylic acid, and 7-diethylaminocoumarin-3-carboxylic acid (Fig. 3.6) were identified as potential donor molecules for effective FRET activity when paired with a flavin acceptor. 7-Diethylaminocoumarin has been extensively used for imaging purposes [13], with maximum absorbance at 410 nm (Fig. 3.5), this dye is well-suited for use with a range of imaging instruments usually equipped with a 405 nm excitation laser.

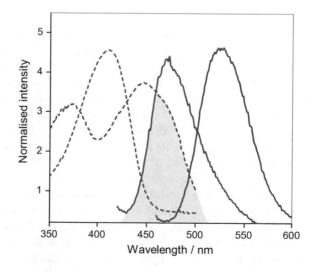

Fig. 3.5 Absorbance (dashed line) and emission (solid line) spectra of 7-diethylaminocoumarin (blue, 10 μM) and N-ethylflavin (red, 10 μM) in HEPES buffer (100 mM, pH 7.4) indicating a significant overlap (shaded yellow) of the emission profile of the coumarin moiety with the absorbance of the flavin

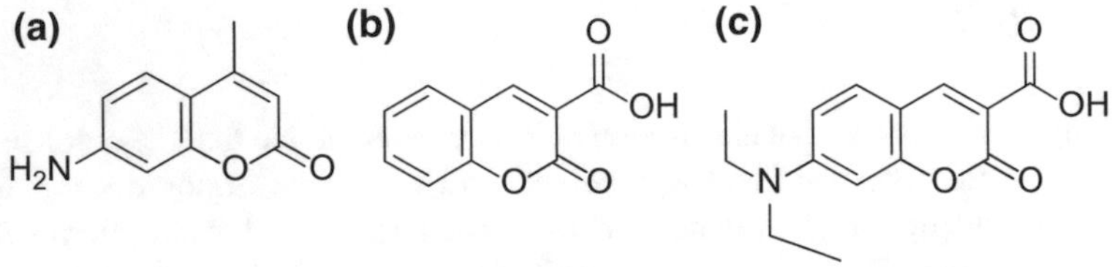

(a)   **(b)**   **(c)**

Fig. 3.6 The chemical structures of **a** 7-amino-4-methyl coumarin, **b** coumarin-3-carboxylic acid and **c** 7-diethylaminocoumarin-3-carboxylic acid

3.2.3 Linker

FRET is a distance-dependent process, where the efficiency of FRET between the donor and acceptor is inversely proportional to the sixth power of the distance between them (Eq. 3.1) [9, 10]. Forster radius (R_o) is the distance at which the efficiency of energy transfer is 50%. For efficient FRET to occur, the donor-acceptor distance (r) must lie within 10–100 Å[14].

$$E = \frac{R_o^6}{(R_o^6 + r^6)} \tag{3.1}$$

E = FRET efficiency
R_o = Forster radius
r = actual donor-acceptor distance

For the purpose of tethering the donor and acceptor fluorophores, short aliphatic hydrocarbon linkers are commonly employed [15]. Furthermore, cycloalkyl linkers with limited rotational freedom, provide greater rigidity and stability to the FRET pair [16, 17]. Therefore, in this study, two types of linkers were utilised: aliphatic hydrocarbon chains 2–3 carbons in length, and 6-membered cycloalkanes bearing suitable functional groups on either side for reaction with flavin and coumarin synthons. Figure 3.7 depicts the structures of target molecules comprising the three vital components to make a FRET-based ratiometric redox probe.

In the oxidised form, the flavin has a planar conformation that is fluorescent, and shows significant overlap of its absorbance with the donor emission profile. It was hypothesised that in the oxidised form, excitation of the coumarin donor would induce a non-emissive energy transfer to the flavin acceptor, resulting in its excitation. The excited flavin would then emit green fluorescence. However, in the reduced form, the flavin molecule assumes a bent conformation that is colourless and non-fluorescent and therefore does not absorb energy in the visible region of the spectrum. As a result, the spectral overlap between the coumarin emission and flavin absorbance is minimised, so excitation of coumarin would result in blue donor emission.

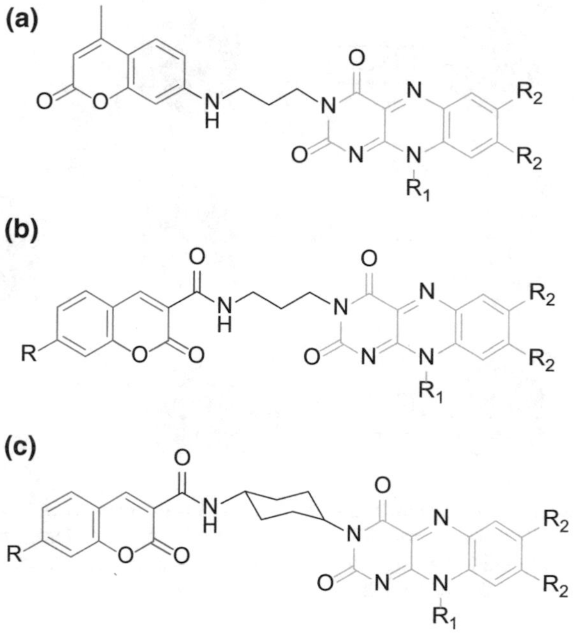

Fig. 3.7 Chemical structures of the target molecules comprising the three components of a FRET-based ratiometric redox probe—the redox responsive flavin group (green), the coumarin FRET-donor (blue) and the linkers (black)

3.3 Synthesis of Flavin Based Ratiometric Redox Probe

Tethering a flavin molecule to a coumarin via an aliphatic linker can essentially be achieved by three synthetic strategies (Fig. 3.8).

1. Installing the linker on the flavin molecule before appending to the coumarin (Fig. 3.8a);
2. Installing the linker on the coumarin molecule before appending to the flavin (Fig. 3.8b);
3. Suitable modification of both the flavin and coumarin moieties to tether them together, thereby generating the linker in a convergent manner (Fig. 3.8c).

All the three strategies were exploited at different stages of this project.

3.3.1 Installing the Linker on the Flavin

The first trial towards the development of a FRET based redox probe involved suitable modification of the flavin molecule in order to prepare it for attachment onto a coumarin, in this case 7-amino-4-methyl coumarin. This would require that the linker on the flavin bears a good leaving group such as a tosyl (*p*-toluenesulfonate) group, for substitution with the amine functional group on the coumarin (Fig. 3.9).

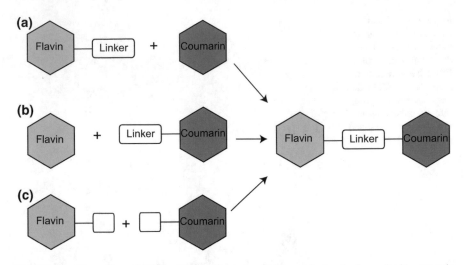

Fig. 3.8 An illustration of different synthetic approaches towards developing a flavin-coumarin FRET probe. **a** Installing the linker on the flavin molecule before appending to the coumarin, **b** installing the linker on the coumarin molecule before appending to the flavin and **c** suitable modification of both the flavin and coumarin moieties to tether them together

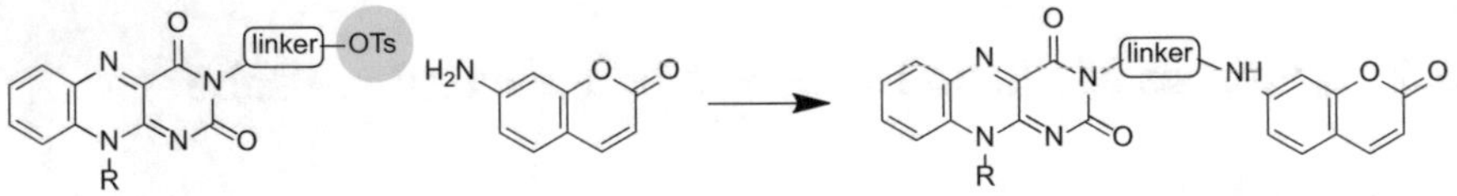

Fig. 3.9 Tosylation strategy to tether flavin and coumarin

3.3.1.1 Synthetic Attempts with Riboflavin

The first flavin investigated was the naturally-occurring riboflavin. However, the polar ribitol tail at the N-10 position of this highly hydrophilic molecule poses synthetic challenges because of its poor solubility in organic solvents. Past attempts to improve the solubility of riboflavin include acetylation of the hydroxyl groups on the ribitol chain, which drastically decreases the polarity of the molecule, making it easier to work with [18].

Acid-catalysed acetylation (Scheme 3.1) of riboflavin (**20**) afforded tetraacetyl-riboflavin (**21**), which was then subjected to alkylation at the N-3 position by 2-bromoethanol in the presence of mild inorganic base to give **22**. **22** bears a terminal -OH group at the N-3 position, and it was hoped that this could then be converted into a tosyl group. Several approaches for tosylation were attempted involving long reaction times, as well as addition of various bases (such as pyridine) both in dilute conditions and neat but, none were successful. It was hypothesised that the acid-base labile acetyl groups might interfere with the tosylation reaction, giving many degradation products.

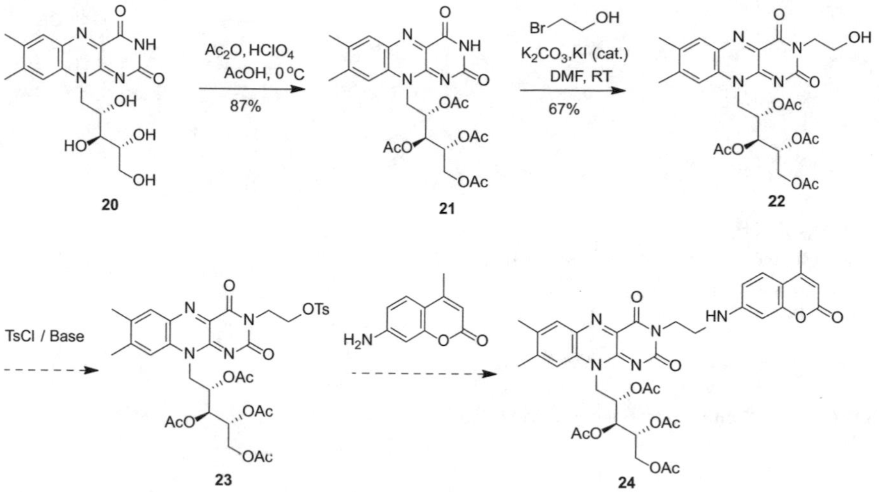

Scheme 3.1 Attempted synthesis of riboflavin-coumarin ratiometric probe **24**

3.3.1.2 Synthetic Attempts with *N*-ethyl flavin

Due to the instability of the acetyl groups present on the tetraacetylriboflavin, synthetic efforts shifted towards *N*-ethyl flavin (**NEF**) (Scheme 3.2), another member of the isoalloxazine family. It was anticipated that *N*-ethyl flavin would have improved solubility compared to riboflavin. Scheme 3.2 shows the steps in the synthesis of *N*-ethyl flavin.

Reaction of 6-chlorouracil (**26**) with excess *N*-ethylaniline (**25**) at high temperature afforded the alkylated product (**27**) [19]. Nitration of **27** with sodium nitrite in acetic acid gave the nitroxide derivative **28** [19], which was then subjected to reduction by sodium dithionite in water, followed by overnight oxidation with hydrogen peroxide to give *N*-ethyl flavin (**NEF**).

NEF was then subjected to similar reactions as for the tosylation strategy of tetraacetylriboflavin outlined in Sect. 3.3.1.1 (Scheme 3.3). Alkylation of *N*-ethyl flavin with 2-bromoethanol gave **29**, but tosylation of the terminal alcohol group could not be achieved despite employing harsh reaction conditions involving the presence of various bases, higher reaction temperatures and longer reaction times.

3.3.2 Installing the Linker on the Coumarin

Unsuccessful attempts with the tosylation strategy shifted attention towards building the linker on the coumarin scaffold such that it bears a suitable functional group for alkylation at the *N*-3 position on the flavin (Fig. 3.8b). For this purpose, coumarin-3-

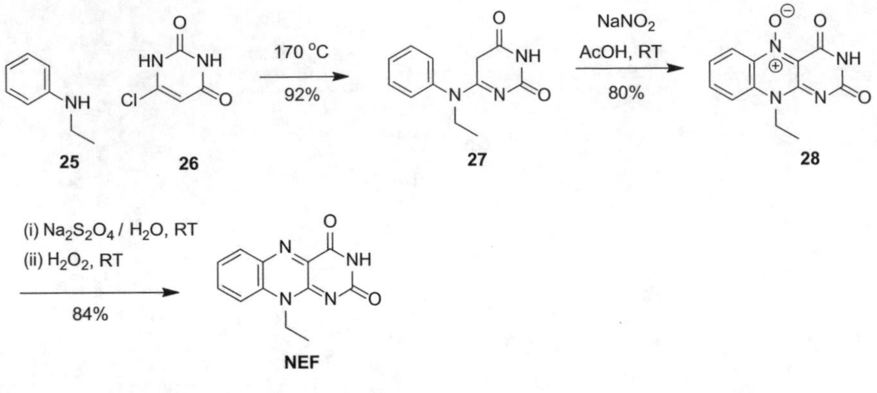

Scheme 3.2 Synthesis of *N*-ethyl flavin (**NEF**)

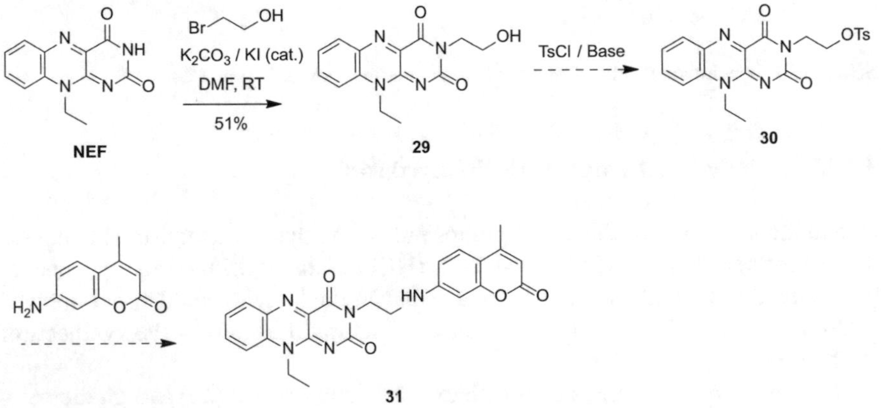

Scheme 3.3 Attempted synthesis of *N*-ethylflavin-coumarin ratiometric probe (**31**)

carboxylic acid was used. Activation of carboxylic acid (**32**) to its *N*-hydroxy succin-
imide ester (**33**) was achieved using the peptide coupling agent dicyclohexylcarbodi-
imide (DCC) (Scheme 3.4). Reaction of **33** with 3-bromopropylamine hydrobromide
in the presence of a base afforded the coumarin amide (**34**) bearing a terminal bromide, which was then successfully utilised for *N*-3 alkylation of **NEF** in the presence
of a base to give flavin coumarin redox sensor (**FCR**).

3.3.2.1 Photophysical and Redox Properties of FCR

Following successful synthesis of **FCR**, its photophysical and redox responsive abilities were tested. Photophysical characterisation **FCR** was performed in HEPES
buffer (100 mM, pH 7.4). In the oxidised form, excitation of **FCR** at 350 nm resulted
in a green fluorescence with maximum emission at 520 nm ($\Phi = 0.233$, Fig. 3.10a).

The absence of the blue donor emission is consistent with an efficient FRET interaction between the two fluorophores in this state. Treatment of **FCR** with a mild reducing agent, sodium cyanoborohydride, caused reduction of the flavin. This resulted in a decrease in the intensity of green fluorescence accompanied by a simultaneous increase in the blue fluorescence band centred at 420 nm ($\Phi = 0.146$, Fig. 3.10a). This is consistent with a decrease in FRET interaction between the donor and acceptor fluorophores. The ratio of flavin to coumarin emission intensities (I_{520}/I_{420}) upon excitation of **FCR** at 350 nm decreased approximately 10-fold upon reduction (Fig. 3.10b).

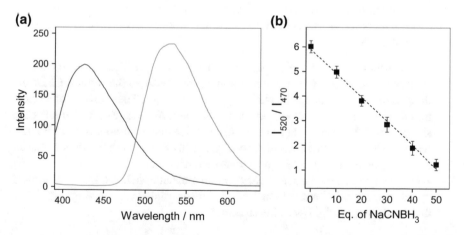

Scheme 3.4 Synthesis of **FCR**

Fig. 3.10 a Fluorescence behaviour of **FCR** (10 µM, $\lambda_{ex} = 350$ nm) in the oxidised (green) and reduced (blue) form upon addition of NaBH$_3$CN, **b** ratio of green/blue emission of **FCR** (10 µM) **b** with incremental addition of NaBH$_3$CN in HEPES buffer (100 mM, pH 7.4)

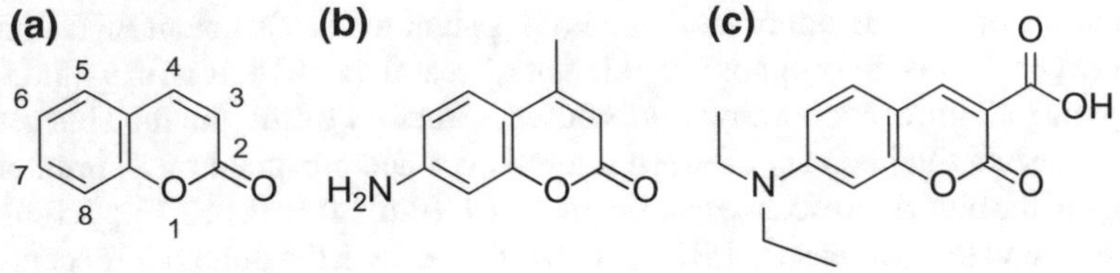

(a)   **(b)**   **(c)**

Fig. 3.11 The chemical structures of **a** coumarin, **b** 7-amino-4-methyl coumarin and **c** 7-diethylaminocoumarin-3-carboxylic acid

3.4 Developing a Second Generation FRET Based Redox Probe

The above results confirmed that coumarin and flavin molecules could be combined to generate a FRET probe, with **FCR** showing a ratiometric fluorescence response to the redox state. Nevertheless, a key shortcoming of **FCR** was its maximum excitation wavelength of 350 nm, which lies in the ultraviolet range. Such short wavelengths of light have been shown to induce the production of ROS/RNS within a biological cell [20]. Using a probe with such short excitation wavelengths could result in false positives, thus limiting the application of **FCR** as a redox probe for biological systems. In addition, exchanging the flexible alkyl linker between the donor and acceptor fluorophores with a constrained one, would improve the FRET efficiency. Therefore, a second generation FRET-based redox probe was designed to incorporate two necessary changes:

1. Bathochromic shift in the excitation and emission profiles of the coumarin, achieved by simple variations in functional groups around the coumarin scaffold that can drastically change its photophysical properties. The presence of an electron donating amine group at position 7 on the molecule (Fig. 3.11a) has been shown to red shift the excitation and emission maxima of coumarins [12]. For this reason, 7-amino-4-methyl coumarin (Fig. 3.11b) and 7-diethylaminocoumarin-3-carboxylic acid (Fig. 3.11c) were considered suitable.
2. A constrained linker based on a hydrocarbon molecule such as a 1,4-disubstituted cyclohexane, or a heterocycle such as a 1,4-disubstituted triazole ring (Fig. 3.12), that would impart rigidity to the structure.

3.4.1 Modification of both Flavin and Coumarin

At this stage, the third strategy mentioned in Sect. 3.3 that involved strategic modification of both flavin and coumarin molecules to generate a linker (Fig. 3.8c), was taken into consideration whilst incorporating the changes outlined above.

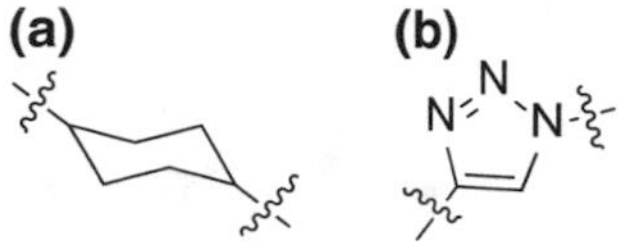

Fig. 3.12 The chemical
structures of constrained
linkers used in this study **a**
1,4-disubstituted cyclohexyl
and **b** 1,4-disubstituted
1,2,3-triazolyl groups

(a)   **(b)**

3.4.1.1 Click Chemistry

Click chemistry is an extensively-utilised synthetic coupling strategy [21]. For
the conjugation of two molecules, the most well-known click reaction is the
copper(I)-catalysed azide-alkyne cycloaddition (CuAAC), which involves a 1,3-
dipolar cycloaddition reaction between a terminal alkyne and an azide in the presence
of Cu(I) to yield a 1,4-disubstituted five membered 1,2,3-triazole ring (Fig. 3.13a)
with very high selectivity. The planar 1,2,3-triazole ring formed as a result of the
click reaction would impart rigidity to the probe structure. It was therefore planned
to use this Cu(I) catalysed click chemistry to develop a second generation FRET
based redox sensor (Fig. 3.13b).

The use of this strategy required an alkyne group on one fluorophore and azide
on the other. Aromatic amines can be readily converted to azides via diazonium
intermediates [22], and by employing this strategy, 7-amino-4-methylcoumarin (**35**)
was converted to a diazonium intermediate, followed by reaction with sodium azide
which yielded the azide building block (**36**) (Scheme 3.5). The alkyne functional
group was then attached to the flavin molecule (**37**) by an *N*-3 alkylation of **NEF**
with propargyl bromide in the presence of a mild inorganic base. The azide (**36**) and
alkyne (**37**) building blocks were then taken forward into a Cu(I) catalysed click
reaction in the presence of Cu(I) iodide and ascorbate, a mild reducing agent used
to prevent oxidation of Cu(I) to Cu(II).

This synthetic strategy was unsuccessful, potentially due to possible reduction
of flavin by the added reducing agent—ascorbate. Carrying out the reaction in the
absence of ascorbate did not result in any reaction progress. TLC analysis showed

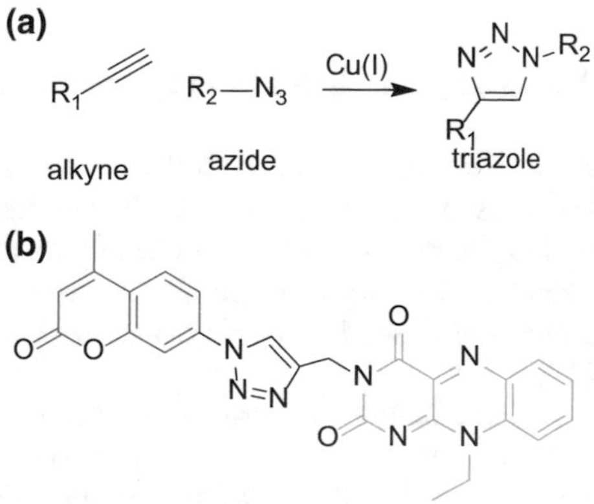

Fig. 3.13 a
Copper(I)-catalysed
azide-alkyne cycloaddition.
b The proposed chemical
structure of the target
molecule depicting the
coumarin donor (blue),
1,4-disubstituted
1,2,3-triazolyl linker (black)
and *N*-ethylflavin (green)

(a)

R_1 ═══ R_2—N_3 $\xrightarrow{Cu(I)}$ triazole

alkyne   azide   triazole

(b)

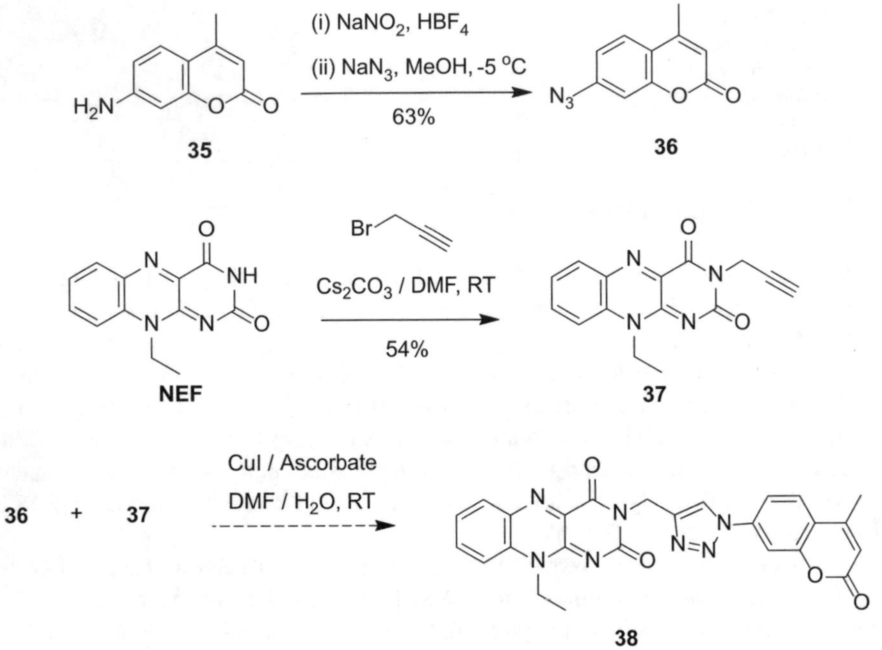

Scheme 3.5 Attempted synthesis of **NEF**-coumarin ratiometric probe via Cu(I) catalysed click reaction

quenching of flavin fluorescence suggesting that Cu(I) was being oxidised to Cu(II) with concomitant reduction of flavin and therefore there was no Cu(I) available to drive the click reaction.

A solution to this undesirable redox chemistry hampering the click reaction could be utilising Ru(II)-based catalysts that could be less susceptible to interference from the redox properties of the flavin [23]. However, given the unavailability of Ru(II) catalyst at that point, other synthetic strategies were attempted.

3.4.1.2   Synthesis of Flavin Coumarin Redox Sensor 1

At this stage, attention reverted to the successful coumarin amide based synthesis as discussed in Sect. 3.3.2; but instead using 7-diethylaminocoumarin-3-carboxylic acid, which has a 60 nm red-shift in excitation maximum compared to coumarin-3-carboxylic acid, and 1,4-*trans*-diaminocyclohexane (Fig. 3.14), a constrained aliphatic hydrocarbon that has been previously reported as a FRET linker [17].

7-diethylaminocoumarin-3-carboxylic acid (**40**) was synthesised by Knovenagel condensation of 4-diethylaminosalicylaldehyde (**39**) with diethylmalonate. **40** was then activated to its *N*-hydroxysuccinimide ester (**41**) using the peptide coupling

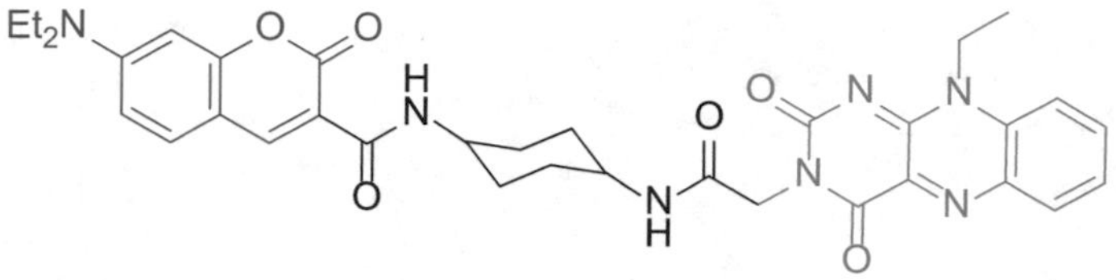

Fig. 3.14 The proposed chemical structure of the target molecule showing the coumarin donor (blue), 1,4-diaminocyclohexane linker (black) and *N*-ethylflavin (green)

agent ethyl diisopropylcarbodiimide (EDC) (Scheme 3.6). Reaction of **41** with *trans*-1,4-diaminocyclohexane, gave the corresponding coumarin amide **42** bearing a reactive terminal amine. **NEF** was synthesised using the standard 6-chlorouracil method [19], giving suitable modification to contain an amine reactive functionality such as activated carboxylic acid. Alkylation of **NEF** with bromoethylacetate in the presence of mild inorganic base resulted in the formation of *N*3-alkylated flavin (**43**).

Condensation of the flavin-based ester **43** with the coumarin amide **42** did not show much reaction progress, possibly due to the two bulky fluorophores. At this stage, generation of a highly reactive carboxylic acid derivative that would readily react with amines such as an acid chloride was considered. To achieve this, the ester **43** was hydrolysed to the carboxylic acid **44**, which was then reacted with neat thionyl chloride (SOCl$_2$) to give the corresponding acid chloride **45** that would readily react with the terminal amine on the coumarin. Condensation of **45** with **42** in the presence of a hindered base DIPEA resulted in a bright yellow solid. The crude solid was then purified by preparative TLC using DCM:methanol (95:5) as the eluent to give flavin coumarin redox sensor 1 (**FCR1**) as a bright orange solid.

3.4.2 Photophysical and Redox Responsive Properties of FCR1

In vitro experiments to characterise the photophysical and redox responsive properties of **FCR1** were performed in HEPES buffer (100 mM, pH 7.4). Considering the overlapping absorbance spectra of coumarin and flavin (Fig. 3.5), an excitation wavelength of 405 nm was chosen as it would result in preferential excitation of the coumarin over flavin. In the oxidised form of **FCR1**, FRET would occur between the coumarin and the flavin, and as a result excitation of **FCR1** with a wavelength of 405 nm resulted in green fluorescence with maximum emission at 525 nm ($\phi = 0.242$) (Fig. 3.15a). Furthermore, the absence of blue donor emission is consistent with an efficient FRET interaction between the two fluorophores in the oxidised state.

As in the case of **FCR**, upon treatment of **FCR1** with a mild reducing agent such as sodium cyanoborohydride, the flavin moiety undergoes reduction. The reduced flavin adopts a bent conformation making it colourless and non-fluorescent [24]. This

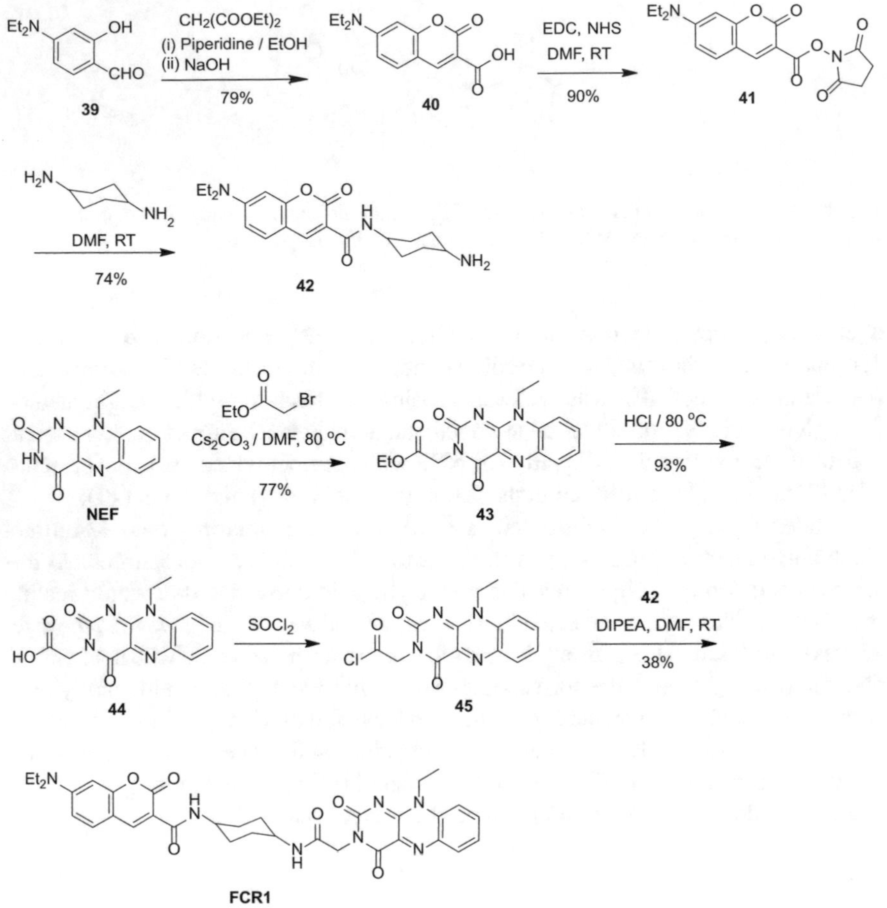

Scheme 3.6 Synthesis of **FCR1**

compromises the ability of the flavin to act as a FRET-acceptor because of minimal spectral overlap between the coumarin and the reduced flavin which results in minimised energy transfer. Consistent with this speculation, upon incremental additions of the reducing agent, **FCR1** exhibited a decrease in the intensity of green fluorescence accompanied by a simultaneous increase in the blue fluorescence band centred at 470 nm ($\Phi = 0.186$). In the case of **FCR1** the overlap between the fluorescence spectra of the reduced and the oxidised form of the probe is larger than for **FCR**, but the non-overlapping areas of the spectra can still be effectively used for quantitative analysis.

Titrating **FCR1** with increasing equivalents of the reducing agent (NaBH$_3$CN) resulted in a linear decrease in the ratio of green-blue emission intensities (I_{520} / I_{470}) upon excitation of **FCR1** at 405 nm (Fig. 3.15a). Complete reduction was achieved

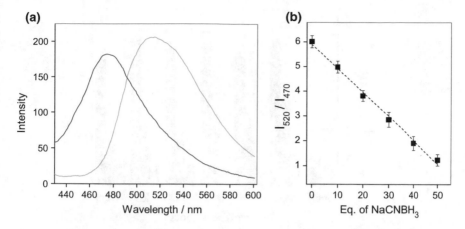

Fig. 3.15 **a** Fluorescence behaviour of **FCR1** (10 μM, $\lambda_{ex} = 405$ nm) in the oxidised (green) and reduced (blue) form upon addition of NaBH$_3$CN, **b** ratio of green/ blue emission of **FCR1** (10 μM) with incremental addition of NaBH$_3$CN in HEPES buffer (100 mM, pH 7.4). Reprinted from *Chemical Communications*, Issue 52, with permission from the Royal Society of Chemistry

with 50 molar equivalents of NaBH$_3$CN resulting in a 6-fold decrease in the ratio (I_{520} / I_{470}) compared to the oxidised form (Fig. 3.15a). These results suggest that the FRET process between the flavin and coumarin remained undisturbed, even when a coumarin with more red-shifted photophysical properties was used. In addition, the presence of another fluorophore molecule (7-diethylaminocoumarin) in close vicinity of the flavin scaffold in **FCR1** did not subvert the redox responsive behaviour of the flavin scaffold.

Re-oxidation of the probe was assessed by reducing **FCR1** (10 μM) completely using NaBH$_3$CN (20 μM) and monitoring the change in the fluorescence ratio I_{520} / I_{470} in the presence of H$_2$O$_2$ (100 μM) (Fig. 3.16). Unlike the other flavin-based redox probes—**NpFR1** and **NpFR2**, which could be re-oxidised completely within 5 min of H$_2$O$_2$ treatment, re-oxidation of **FCR1** was observed to be slower, requiring about 60 minutes. Similar results were obtained from re-oxidation of reduced **FCR1** (10 μM **FCR1** in the presence of 20 μM NaBH$_3$CN) in the presence of various biologically relevant ROS/RNS (100 μM) produced in situ (Fig. 3.16). Although slow, **FCR1** could be re-oxidised by all the ROS/RNS, with complete re-oxidation within 60 min, but such long re-oxidation times limit the use of **FCR1** as a reversible redox sensor. It is suggested that the slow re-oxidation in case of **FCR1** might be a consequence of its reduced form being relatively more stable. Electrochemical experiments were performed to investigate this hypothesis.

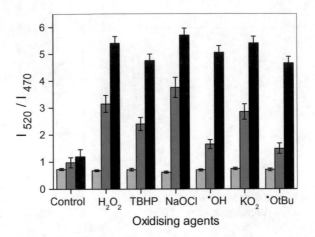

Fig. 3.16 Oxidation of **FCR1** with various oxidising agents. Bars represent the ratio of green to blue fluorescence intensity (520/ 470 nm, λ_{ex} = 405 nm) after reduction (red) of **FCR1** (10 μM in the presence of 20 μM NaBH$_3$CN) followed by re-oxidation of reduced **FCR1** 30 min (blue) and 60 min after (black) the addition of 100 μM oxidising agent. Error bars represent standard deviation n = 3. Reprinted from *Chemical Communications*, Issue 52, with permission from the Royal Society of Chemistry

3.4.3 Electrochemical Studies

The redox responsive range of **FCR1** was determined from electrochemical studies. Cyclic voltammograms were recorded using a 2 mM solution of **FCR1** in a freshly distilled acetonitrile containing 0.1 M tetrabutylammonium hexafluorophosphate (TBAPF$_6$) as a supporting electrolyte and 1 mM Fc/Fc$^+$ couple as an internal standard. As seen in Fig. 3.17, the shape of the cyclic volatmmogram confirmed that the reduction-oxidation events of the probe are reversible and the reduction potential was calculated to be −1.15 V (vs Fc/Fc$^+$), which is similar to the value of reduction potential of riboflavin −1.18 V (vs Fc/Fc$^+$) reported in the literature [25]. This confirmed that the reduction potential of **FCR1** lies within the biologically pertinent window, and tethering a second fluorophore did not alter the redox potential of the flavin component.

In cyclic voltammetry, the Randles-Sevcik equation (Eq. 3.2) explains the effect of scan rate on the peak current, and the slope of a plot between I_{pc} versus the square root of the scan rate will be proportional to the diffusion coefficient [26]. For a reversible system the peak current varies linearly with the square root of the scan rate. To investigate if the electrochemistry of **FCR1** followed the Randles-Sevcik equation, cyclic voltammograms of **FCR1** (2 mM in MeCN with 0.1 M TBAPF$_6$ and ferrocene) were recorded at various scan rates ranging from 0.02 to 0.7 Vs^{-1}. The asymmetry of the cathodic and anodic branches suggests a degree of chemical irreversibility of the molecule, which also explains the slow re-oxidation (Fig. 3.18a).

Fig. 3.17 Cyclic voltammogram showing the reduction and oxidation events of **FCR1** (2 mM) in MeCN containing 0.1 M TBAPF$_6$ as supporting electrolyte. Two consecutive redox cycles were recorded and no variations were observed between the cycles. Reprinted from *Chemical Communications*, Issue 52, with permission from the Royal Society of Chemistry

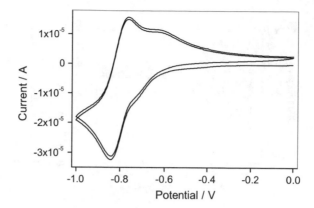

$$I_p = 269n^{\frac{3}{2}}AD^{\frac{1}{2}}Cv^{\frac{1}{2}} \qquad (3.2)$$

I_p = peak current (amp)
n = number of electrons
A = electrode area (cm^2)
D = diffusion coefficient (cm^{2}s^{-1})
C = concentration (molL^{-1})
v = scan rate (Vs^{-1})

Furthermore, the cathodic peak current (reduction process) varies linearly with the square root of scan rate, showing that the electrochemical reduction of **FCR1** is a normal diffusion-controlled process under these conditions (Fig. 3.18b). In order to better understand the effect of electrochemical reduction and oxidation on the photophysical properties of **FCR1**, spectro-electrochemical studies were performed.

3.4.4 Spectro-Electrochemistry

Spectro-electrochemistry is a hybrid technique developed by combining electro-chemistry and spectroscopy. Here, the oxidation/reduction of a molecule is brought about electrochemically by the addition/removal of electrons at an electrode, while the solution around the electrode is simultaneously interrogated spectroscopically. This is a convenient technique that allows to record information on the photophysi-cal properties and redox potentials simultaneously, and enables understanding of the spectroscopic properties of electrochemically-generated species.

Spectro-electrochemical measurements were performed in Dr Conor F. Hogan's lab at the La Trobe University. In these experiments, platinum gauze was used as the working electrode, platinum wire as the counter electrode and Ag/Ag$^+$ as non-aqueous reference electrode. A reduction potential of -1.3 V was applied, and the

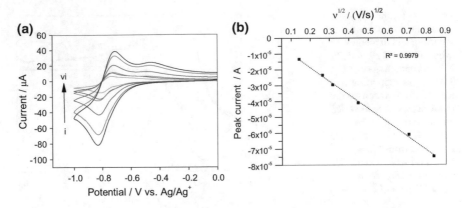

Fig. 3.18 a Cyclic voltammograms for **FCR1** (2 mM) at various scan rates in MeCN containing 0.1 M TBAPF$_6$ as supporting electrolyte. The working electrode was a 3 mm diameter glassy carbon electrode and the scan rates were: (i) 0.02, (ii) 0.07, (iii) 0.1, (iv) 0.2, (v) 0.5 and (vi) 0.7 Vs^{-1}. **b** Plot of peak current versus square root of scan rate for the voltammetric reduction of **FCR1** (2 mM) in MeCN. Reprinted from *Chemical Communications*, Issue 52, with permission from the Royal Society of Chemistry

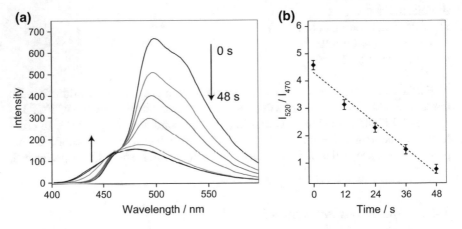

Fig. 3.19 a Fluorescence behaviour and **b** ratio of green/blue emission of **FCR1** (100 μM) over time after the application of a potential of −1.3 V. Reprinted from *Chemical Communications*, Issue 52, with permission from the Royal Society of Chemistry

fluorescence spectra were recorded at 12 s intervals using a 405 nm excitation. Over subsequent scans, a decrease in the green fluorescence ($\lambda_{max} = 525$ nm) could be observed, with a corresponding increase in the peak at 470 nm, but this increase is much smaller in magnitude compared to that observed in the case of chemical reduction (Fig. 3.19a).

As suggested by the voltammetric analysis in Fig. 3.18, the product of the electrochemical reduction seems to be less stable in acetonitrile compared to aqueous media (HEPES buffer), but the excellent stability of acetonitrile under both oxidising and reducing conditions [26] demanded its use. Despite this, Fig. 3.19b shows that the change in the fluorescence ratio I_{520}/I_{470} followed a trend similar to that observed in case of chemical reduction.

Thus, the electrochemical and spectro-electrochemical studies confirmed that **FCR1** is a reversible redox sensor with its reduction potential well within the biological window. Furthermore, ratiometric response of **FCR1** is maintained by both chemical and electrochemical reduction.

3.4.5 Control Experiments

The current shift towards a preference for ratiometric probes has arisen primarily from the fact that the ratiometric response remains unaltered by other chemical species in the probe environment as well as other factors such as variations in the instrumental optics or probe concentration. Although **FCR** established the flavin-coumarin FRET pair as a ratiometric redox sensor, its lower excitation maximum limits its use in biological systems. Therefore, further studies were performed using the more biologically-suitable **FCR1**. In order to assess the applicability of **FCR1** as a ratiometric fluorescent probe in biological systems, control experiments were carried out to assess any effects that pH or biological concentrations of transition metals might have on the ratiometric response of the probe.

The fluorescence ratio (I_{520} / I_{470}) of **FCR1** did not change significantly across the range of biologically relevant pH values (4–9; Fig. 3.20a). Furthermore, as depicted in Fig. 3.20b, the ratiometric response of the probe remained unaltered in the presence of 50 equivalents of common metal ions. These results suggest that other aspects of the

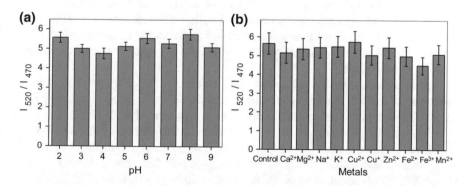

Fig. 3.20 The ratio of fluorescence emission from **FCR1** (10 μM) **a** over a range of pH values and **b** in the presence of common metal ions (100 μM). Bars represent the ratio of the fluorescence intensity (520/470 nm, $\lambda_{ex} = 405$ nm) as the mean of three replicates. Error bars represent standard deviation, $n = 3$. Reprinted from *Chemical Communications*, Issue 52, with permission from the Royal Society of Chemistry

biological environment minimally interfere with probe response, thus highlighting the robustness of **FCR1** and its applicability for biological imaging.

3.4.6   Confocal Microscopy

Prior to performing any biological experiments it was essential to ensure that **FCR1** did not exert any cytotoxic effects. A standard MTT cell viability assay was carried out by treating HeLa cells with concentrations of **FCR1** ranging from 0–160 μM for 24 h. The value of IC_{50} was found to be 80 μM for 24 h. Therefore establishing the upper limits of the duration and dose of the probe that can be applied to biological systems for sensing applications.

The applicability of **FCR1** to redox imaging of cultured cells was then assessed. When compared with the confocal microscope, multi-photon microscopy has a number of advantages, such as reduced photo-toxicity to cells, minimal photo-bleaching of the dyes and better tissue penetration [27]. Therefore, imaging experiments were performed using a multi-photon confocal microscope. A series of two-photon exci- tation wavelengths were tested to identify the wavelengths that gave the highest signal-to-noise (S/N) ratio. Each excitation wavelength was used to obtain three images of HeLa cervical cancer cells treated with **FCR1** (10 μM, 15 min, $\lambda_{em} =$ 500–600 nm). Analysis of the collected images showed that the highest signal to noise ratio (S/N) was obtained from a two photon excitation wavelength of 820 nm (Table 3.1). This was therefore chosen to be the excitation wavelength for further biological investigations.

Furthermore, it was essential to confirm that the photophysical properties of the oxidised and reduced forms of **FCR1** at two-photon excitation wavelength of 820 nm are not different to those observed at a single-photon excitation of 405 nm. This was achieved by performing a spectral scan of solutions of reduced and oxidised form of **FCR1** using a microscope equipped with 405 nm and multi-photon lasers. Figure 3.21 confirms that the spectroscopic behaviour of the reduced and oxidised forms of **FCR1** at 405 nm single photon and 820 nm two-photon excitation was not significantly different.

Having demonstrated the redox sensitivity, ratiometric response and non-cytotoxic behaviour of **FCR1**, the next step was to test its ability to respond to changes in the oxidative capacity of cultured cells. HeLa cells treated with **FCR1** (10 μM, 15 min) and excited using a two-photon excitation wavelength of 820 nm showed significant fluorescence in both blue (420–470 nm) and green channels (520–600 nm), while untreated control cells showed no appreciable fluorescence in either region (Fig. 3.22). The concentration of **FCR1** used in these experiments is far below the IC_{50} value calculated from the MTT cell viability assay, ensuring that cytotoxic effects did not effect the validity of the experiment.

In order to mimic the conditions of resting cells, as well as reducing and oxidising cell environments, HeLa cells treated with **FCR1** (10 μM, 15 min) were further subjected to treatment with the vehicle control (PBS, 50 μM, 30 min), the reduc-

Table 3.1 Signal to noise ratios for a range of two photon excitation wavelengths. Signal to noise ratios are an average of 3 images acquisitions

Wavelength / nm	Trans / %	Gain / %	Offset / %	S/N ratio
700	64	73	42	240.3
720	35	73	45	251.9
740	22	73	47	243.7
760	22	73	50	249.8
780	27	73	52	261.5
800	33	73	54	337.4
820	43	73	57	359.8
840	50	73	60	317.6
860	50	73	63	210.2
880	56	73	67	168.3
900	49	73	69	171.2
920	68	78	70	149.5
940	79	78	72	104.6
960	88	80	76	74.9
980	97	85	71	98.4

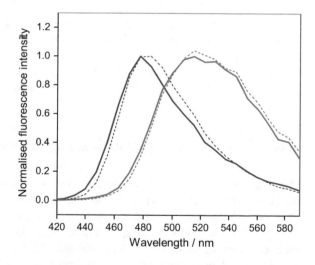

Fig. 3.21 Fluorescent spectra of **FCR1** (2 μM) oxidised (green) and reduced form (blue) obtained following excitation at 405 nm (dashed) and 820 nm (solid). Reduction was achieved by the addition of NaBH$_3$CN (100 μM). Reprinted from *Chemical Communications*, Issue 52, with permission from the Royal Society of Chemistry

tant *N*-acetyl cysteine (NAC, 50 μM, 30 min) or the oxidant H$_2$O$_2$ (50 μM, 30 min) (Fig. 3.23). The cells were then visualised to obtain images in both blue and green channels. A FIJI plugin, RatioPlus, was then employed to obtain ratio images by dividing the green channel image with the blue one, and the obtained images were pseudo-coloured to indicate the ratio of emission intensities. The reduced cells demonstrated a lower intensity ratio when compared to cells treated with probe alone,

Fig. 3.22 Two-photon confocal microscopy imaging of HeLa cells treated with vehicle control (DMSO) and **FCR1** (10 μM, 15 min, $\lambda_{ex} = 820$ nm) in blue and green channels. Scale bar represents 20 μm. Reprinted from *Chemical Communications*, Issue 52, with permission from the Royal Society of Chemistry

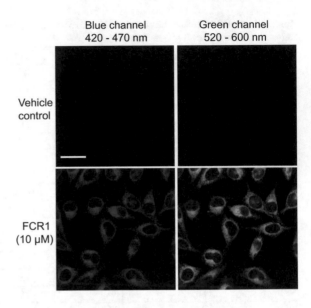

while the oxidised cells showed a higher ratio, in agreement with the FRET-based ratio changes in solution (Fig. 3.23).

In addition, the oxidative capacity of peroxide-treated HeLa cells was analysed at different time points (Fig. 3.24). These studies showed that, with increasing peroxide treatment times up to 1 h, there was an increase in the average intensity ratio of **FCR1** and thereby the cells' oxidative capacity. A much lower intensity ratio after 2 h of peroxide treatment highlights the cells' ability to restore its redox homoeostasis.

3.4.7 Fluorescence Lifetime Imaging Microscopy

Fluorescence properties are not limited to the spatial distribution of the fluorescence intensity and the fluorescence spectrum. Another characteristic gaining increasing attention is fluorescence lifetime, which depends on the fluorescence decay function [28]. For a homogeneous population of molecules, the fluorescence decay of a molecule from the excited state to ground state is a single exponential function [9]. The time constant of this function is the fluorescence lifetime, which typically lies in the picosecond to nanosecond time scale [9]. The fluorescence lifetime depends on the nature of the molecule, its conformation and interactions with surrounding environment [9].

Fluorescence lifetime imaging microscopy (FLIM) techniques can be broadly classified into time-domain and frequency domain protocols. Time correlated single photon counting (TCSPC) is most widely used owing to the high time resolution, short acquisition times, improved lifetime accuracy and photon counting efficiency

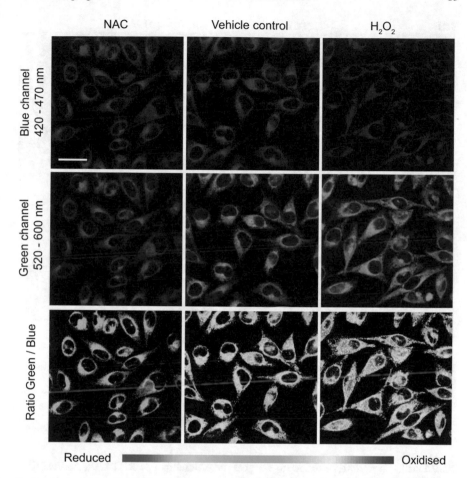

Fig. 3.23 Two-photon confocal microscopy imaging of HeLa cells treated with **FCR1** (10 μM, 15 min, $\lambda_{ex} = 820$ nm) and **a** N-acetyl cysteine (50 μM, 30 min), **b** vehicle control and **c** H_2O_2 (50 μM, 30 min) in blue and green channels. The pseudo colour ratio images indicate the ratio of emission intensity in the green channel to blue channel. Scale bar represents 20 μm. Reprinted from *Chemical Communications*, Issue 52, with permission from the Royal Society of Chemistry

that can be obtained using this technique and was the technique employed in this study [29]. TCSPC FLIM involves exciting the fluorophore by short laser pulses, followed by measuring the arrival times of single photons of the fluorescence at the detector with respect to the laser pulses and the position of the laser beam (Fig. 3.25) [29]. A histogram is generated, depicting an exponential decay of the frequency of photons over time. The decay profile is then deconvoluted and subjected to fitting procedure to calculate the fluorescence lifetime [29].

Fluorescence lifetime imaging microscopy (FLIM) has been employed in a variety of applications such as resolving overlapping spectra, sensing environmental factors such as pH, viscosity and proximity or binding of fluorophore to a biomolecule [28].

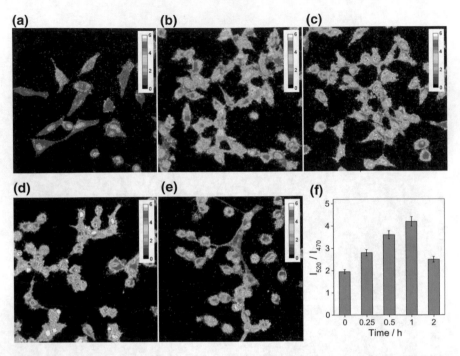

Fig. 3.24 Pseudo-coloured ratio images (green/blue) of HeLa cells treated with H_2O_2 (50 μM) for **a** 0 h, **b** 0.25 h, **c** 0.5 h, **d** 1 h and **e** 2 h, followed by treatment with **FCR1** (10 μM) for 15 mins. Average green/blue ratio of individual cells after different durations of H_2O_2 treatment (**f**) are also indicated. Error bars indicate standard deviation $n = 3$. Reprinted from *Chemical Communications*, Issue 52, with permission from the Royal Society of Chemistry

FLIM is also frequently used to investigate the efficiency of FRET [29]. As described in Sect. 3.1.1, FRET is an interaction between two fluorophores, wherein the emission spectrum of one (donor) overlaps with the absorption spectrum of the other (acceptor), resulting in a non-radiative energy transfer from the donor fluorophore in its excited state to the acceptor fluorophore. Therefore one consequence of FRET is the quenching of the donor fluorescence resulting in a decrease in the donor lifetime (Fig. 3.26) [9]. FRET efficiency can be calculated by comparing the fluorescence lifetimes of the donor in the presence and absence of acceptor molecule using the formula [29]

$$E = 1 - \frac{\tau_{DA}}{\tau_D} \tag{3.3}$$

E = FRET efficiency
τ_{DA} = mean lifetime of the donor in the presence of acceptor
τ_D = mean lifetime of the donor alone

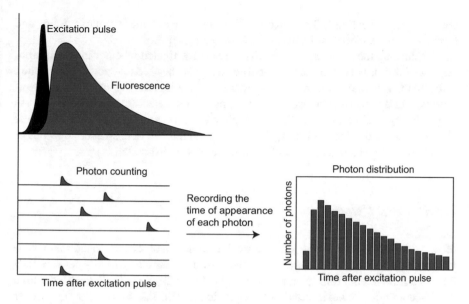

Fig. 3.25 Operation principle of time-correlated single photon counting (TCSPC) measurements. The sample is excited by a pulsed laser source and the photons emitted are detected with a high-gain photomultiplier and the time with respect to the excitation pulse is measured. A histogram is then generated and the photon distribution over time is built. (Figure reproduced from: Becker, The bh TCSPC Handbook [29])

Fig. 3.26 Influence of FRET on the donor decay lifetime functions. In the absence of FRET the unquenched donor usually exhibits a mono-exponential decay profile, whereas in the presence of FRET the quenched donor exhibits a bi-exponential decay profile

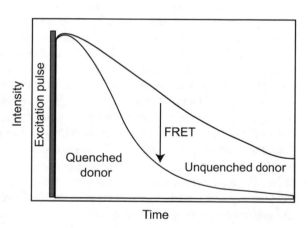

The FRET efficiency of **FCR1** was examined by FLIM, investigating HeLa cells treated with either the 7-diethylaminocoumarin donor alone or with **FCR1**. In the absence of acceptor, the donor lifetimes were found to fit a single component decay curve, with lifetimes of 2.3 ns (Fig. 3.27a), whilst in the presence of the acceptor, a two component fit with lifetimes of 1.1 ns (69%) and 2.3 ns (31%) was obtained

(Fig. 3.27b), indicating a 36% FRET efficiency between the coumarin donor and flavin acceptor moieties in **FCR1**, calculated using Eq. 3.3.

Considering the advantages of FLIM, it was investigated whether the mean fluorescence lifetimes of the donor fluorophore could also be used to report cellular redox state. To measure the fluorescence lifetimes of **FCR1** in reducing and oxidising conditions, HeLa cells were prepared in the same way as mentioned in Sect. 3.4.6. As shown in Fig. 3.28, reduced cells have higher T_m (2.0 ns) in comparison to normal (1.7 ns) and oxidised cells (1.3 ns). This suggested that the fluorescence lifetimes of **FCR1** could also be used as an indicator of cellular redox state.

3.4.8 Flow Cytometry

Flow cytometry is an imaging technology that simultaneously interrogates and analyses multiple physical and chemical characteristics of particles, usually cells, as they flow in a fluid stream through a beam of light (Fig. 3.29) [30]. Cells labelled with a fluorescent molecule are inserted as a suspension into the fluidics of a flow cytometer which then focuses the cells into a stream of droplets each containing a single cell. A laser beam then interrogates this stream of cells, one cell at a time and the fluorescence from each cell is recorded. Thousands of cells per second can be analysed and the data corresponding to the entire cell population can be analysed.

Recently there has been great interest in the development and application of ratiometric probes for use in flow cytometry experiments, such as FURA Red, a ratiometric calcium sensor [31], and JC-1, a ratiometric probe for mitochondrial membrane potential [32]. To investigate the effects of the reducing and oxidising agents (at concentrations used thus far) on a large population of cells, and the ability of **FCR1** (10 µM, 15 min) to report on these changes, flow cytometry experiments were performed. HeLa cells treated with **FCR1** were interrogated by flow cytometry with excitation at 405 nm, and emission collected with windows centred at 450/50 and 560/35 nm. Cells treated with **FCR1** showed considerably greater fluorescence in both windows than untreated cells (Fig. 3.30a). The fluorescence ratio of the population of cells treated with H_2O_2 was considerably higher than that of control cells, and the ratio of NAC-treated population was much lower (Fig. 3.30b). These results demonstrate the diversity of **FCR1** in terms of its compatibility with different modalities which can applied for investigating redox state.

3.5 Conclusions

FCR1 shows great promise as a tool to study oxidative stress in biology. With its ratiometric output, the probe can be used to observe changes in oxidative capacity without interference from background effects such as probe concentration and other environmental factors. Furthermore, the design of **FCR1** also validates the use of a

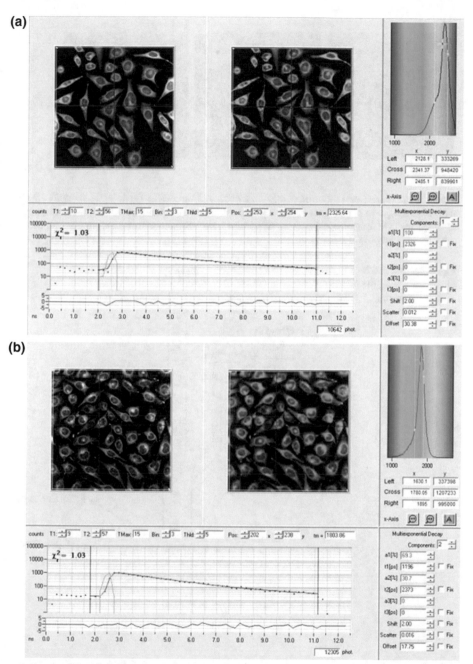

Fig. 3.27 Analysis of donor fluorescence lifetimes (420–470 nm) of HeLa cells treated with **a** donor only (10 μM, 15 min, $\lambda_{ex} = 820$ nm) and **b FCR1** (10 μM, 15 min, $\lambda_{ex} = 820$ nm) indicating a single component fit with a lifetime of 2.3 ns. Pixel by pixel lifetime decay map has been colour coded from 0.9 to 2.7 ns. Reprinted from *Chemical Communications*, Issue 52, with permission from the Royal Society of Chemistry

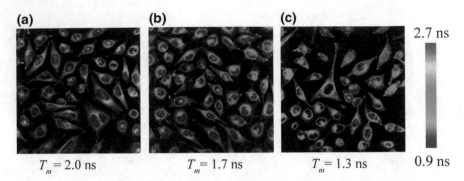

(a)   **(b)**   **(c)**

$T_m = 2.0$ ns $T_m = 1.7$ ns $T_m = 1.3$ ns

2.7 ns

0.9 ns

Fig. 3.28 **a** Fluorescence lifetimes of the donor (420–470 nm) in HeLa cells treated with **FCR1** (10 μM, 15 min, $\lambda_{ex} = 820$ nm) and N-acetyl cysteine, vehicle control and H_2O_2. Pseudo-colour images represent mean lifetime. Reprinted from *Chemical Communications*, Issue 52, with permission from the Royal Society of Chemistry

Fig. 3.29 Schematic diagram showing the process of flow cytometry

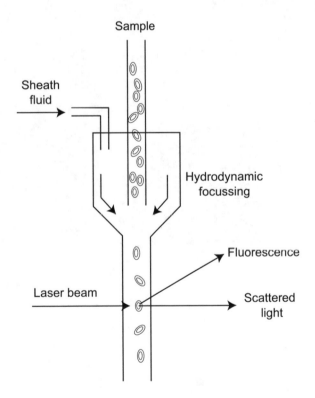

Sample

Sheath fluid

Hydrodynamic focussing

Fluorescence

Laser beam

Scattered light

FRET strategy for developing ratiometric probes. FRET could therefore be employed to develop ratiometric versions of commercial redox probes which currently possess excellent intensity-based readout for biological redox state. Re-oxidation of **FCR1** could be achieved with all the cellular ROS, indicating its applicability as a probe

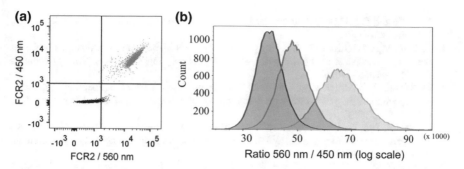

Fig. 3.30 Flow cytometric studies of HeLa cells treated with **FCR1** (10 μM, 15 min, $\lambda_{ex} = 820$ nm) **a** Dot plots show that cells treated with **FCR1** (red) have higher emission intensities in green (560 nm) and blue (450 nm) channels from than those treated with vehicle control (black). **b** Histograms showing the fluorescence ratio (560/450 nm) of cells treated with N-acetyl cysteine (blue), vehicle control (green) and H_2O_2 (red). Reprinted from *Chemical Communications*, Issue 52, with permission from the Royal Society of Chemistry

for a global cellular oxidative capacity. While the slow re-oxidation of **FCR1** limits its use as a reporter for redox fluxes within a cell, nevertheless, **FCR1** is the first ratiometric probe based on a flavin molecule and further interrogations would be necessary to achieve a prompt response towards re-oxidation.

Furthermore, examination of the electrochemical properties of the probe indicates that the modifications made around the flavin scaffold did not alter the redox responsive behaviour of **FCR1** whilst confirming its reduction potential is biologically relevant. The asymmetry in the cathodic and anodic events suggest a quasi-reversible reduction of **FCR1**. Spectro-electrochemistry ensured that a consistent ratiometric behaviour was observed whether the reduction of **FCR1** was induced chemically or electrochemically.

The preliminary biological experiments detailed here validate the ability of **FCR1** to act as a ratiometric reporter of cellular oxidative capacity in reduced and oxidised cells. As well as demonstrating the utility of **FCR1** in detecting changes by confocal microscopy, it is evident that **FCR1** is a useful tool to investigate changes in oxidative capacity by employing other modalities such as FLIM and flow cytometry. This probe was further employed to study more complex biological model systems, which is discussed in Chaps. 6 and 8.

References

1. X. Wang, W.-X. Qi, Y.-Q. Xia, B. Tang, Progress on fluorescent probes for reversible redox cycles and their application in living cell imaging. Chin. J. Anal. Chem. **40**, 1301–1308 (2012)
2. A. Kaur, J.L. Kolanowski, E.J. New, Reversible fluorescent probes for biological redox states. Angew. Chem. Int. Ed. Engl. **55**, 1602–13 (2016)

3. K.M. Fisher, C.J. Campbell, Ratiometric biological nanosensors. Biochem. Soc. Trans. **42**, 899–904 (2014)
4. A. Kaur, M.A. Haghighatbin, C.F. Hogan, E.J. New, A FRET-based ratiometric redox probe for detecting oxidative stress by confocal microscopy. FLIM Flow Cytometry. Chem. Commun. **51**, 10510–10513 (2015)
5. D. Srikun, E.W. Miller, D.W. Domaille, C.J. Chang, An ICT-based approach to ratiometric fluorescence imaging of hydrogen peroxide produced in living cells. J. Am. Chem. Soc. **130**, 4596–4597 (2008)
6. J. Fan, W. Sun, M. Hu, J. Cao, G. Cheng, H. Dong, K. Song, Y. Liu, S. Sun, X. Peng, An ICT-based ratiometric probe for hydrazine and its application in live cells. Chem. Commun. **48**, 8117–8119 (2012)
7. B. Liu, H. Wang, T. Wang, Y. Bao, F. Du, J. Tian, Q. Li, R. Bai, A new ratiometric ESIPT sensor for detection of palladium species in aqueous solution. Chem. Commun. **48**, 2867–2869 (2012)
8. Z. Song, R.T.K. Kwok, E. Zhao, Z. He, Y. Hong, J.W.Y. Lam, B. Liu, B.Z. Tang, A ratiometric fluorescent probe based on ESIPT and AIE processes for alkaline phosphatase activity assay and visualization in living cells. ACS Appl. Mater. Interfaces **6**, 17245–17254 (2014)
9. J. Lakowicz, *Principles of Fluorescence Spectroscopy* (Kluwer Academic/Plenum Publishers, New York, Boston, Dordrecht, London, Moscow, 1999)
10. E.A. Jares-Erijman, T.M. Jovin, FRET imaging. Nat. Biotechnol. **21**, 1387–1395 (2003)
11. L.D. Lavis, T.-Y. Chao, R.T. Raines, Latent blue and red fluorophores based on the trimethyl lock. ChemBioChem **7**, 1151–1154 (2006)
12. N.A. Kuznetsova, O.L. Kaliya, The photochemistry of coumarins. Russian Chem. Rev. **61**, 683 (1992)
13. L. Yuan, W. Lin, K. Zheng, S. Zhu, FRET-based small-molecule fluorescent probes: rational design and bioimaging applications. Acc. Chem. Res. **46**, 1462–1473 (2013)
14. R. Roy, S. Hohng, T. Ha, A practical guide to single-molecule FRET. Nat. Meth. **5**, 507–516 (2008)
15. D.J. Crawford, A.A. Hoskins, L.J. Friedman, J. Gelles, M.J. Moore, Single-molecule colocalization FRET evidence that spliceosome activation precedes stable approach of 5 splice site and branch site. Proc. Natl. Acad. Sci. **110**, 6783–6788 (2013)
16. M.C. Morris, Fluorescence-based biosensors: from concepts to applications, in *Progress in Molecular Biology and Translational Science* (Elsevier Science, 2012)
17. A.E. Albers, V.S. Okreglak, C.J. Chang, A FRET-based approach to ratiometric fluorescence detection of hydrogen peroxide. J. Am. Chem. Soc. **128**, 9640–9641 (2006)
18. F. Muller, *NMR Spectroscopy on Flavins and Flavoproteins* (2014)
19. F. Yoneda, Y. Sakuma, M. Ichiba, K. Shinomura, Syntheses of isoalloxazines and isoalloxazine 5-oxides a new synthesis of riboflavin. J. Am. Chem. Soc. **98**, 830–835 (1976)
20. C.-H. Lee, S.-B. Wu, C.-H. Hong, H.-S. Yu, Y.-H. Wei, Molecular mechanisms of UV-induced apoptosis and its effects on skin residential cells: the implication in UV-based phototherapy. Int. J. Mol. Sci. **14**, 6414–35 (2013)
21. H.C. Kolb, M.G. Finn, K.B. Sharpless, Click chemistry: diverse chemical function from a few good reactions. Angew. Chem. Int. Ed. **40**, 2004–2021 (2001)
22. S. Bräse, C. Gil, K. Knepper, V. Zimmermann, Organic azides: an exploding diversity of a unique class of compounds. Angew. Chem. Int. Ed. **44**, 5188–5240 (2005)
23. L. Zhang, X. Chen, P. Xue, H.H.Y. Sun, I.D. Williams, K.B. Sharpless, V.V. Fokin, G. Jia, Ruthenium-catalyzed cycloaddition of alkynes and organic azides. J. Am. Chem. Soc. **127**, 15998–15999 (2005)
24. A.J.W.G. Visser, S. Ghisla, V. Massey, F. MÜLler, C. Veeger, Fluorescence properties of reduced flavins and flavoproteins. Eur. J. Biochem. **101**, 13–21 (1979)
25. B. König, M. Pelka, H. Zieg, T. Ritter, H. Bouas-Laurent, R. Bonneau, J.-P. Desvergne, Photoinduced electron transfer in a phenothiazine riboflavin dyad assembled by zincimide coordination in water. J. Am. Chem. Soc. **121**, 1681–1687 (1999)
26. C.G. Zoski, *Handbook of Electrochemistry* (Elsevier, 2007)

27. P.T.C. So, C.Y. Dong, B.R. Masters, K.M. Berland, Two-photon excitation fluorescence microscopy. Annu. Rev. Biomed. Eng. **2**, 399–429 (2000)
28. M.Y. Berezin, S. Achilefu, Fluorescence lifetime measurements and biological imaging. Chem. Rev. **110**, 2641–2684 (2010)
29. W. Becker, *The Bh TCSPC Handbook: Time-correlated Single Photon Counting Modules SPC-130, SPC-134, SPC-130 EM, SPC-134 EM, SPC-140, SPC-144, SPC-150, SPC-154, SPC-630, SPC-730, SPC-830; Simple-Tau Systems, SPCM Software* (SPCImage Data Analysis, Becker et Hickl, 2012)
30. H.M. Shapiro, *Practical Flow Cytometry* (Wiley, 2005)
31. E.R. Wendt, H. Ferry, D.R. Greaves, S. Keshav, Ratiometric analysis of fura red by flow cytometry: a technique for monitoring intracellular calcium flux in primary cell subsets. PLoS ONE **10**, e0119532 (2015)
32. A. Cossarizza, S. Salvioli, Flow cytometric analysis of mitochondrial membrane potential using JC-1, in *Current Protocols in Cytometry*, ed. by J. Paul Robinson, et al., Chap. 9 (2001), Unit 9.14

Chapter 4
Mitochondrially-Targeted Ratiometric Redox Probes

The oxidative capacity and levels of ROS production throughout the cell are by no means homogeneous, and mitochondria are the cardinal players in cellular redox homoeostasis and signalling. Mitochondrial ROS levels are known to be key to the function of the organelle, particularly in redox signalling processes, which have a variety of physiological roles, including the maintenance of mitochondrial morphology [1, 2], stem cell differentiation [3] and cardiac remodelling [4]. On the other hand, mitochondrial oxidative stress is implicated in diseases associated with ageing [5, 6].

Despite the vast interest in elucidating the role of mitochondrial redox state in cellular signalling and disease, there is a paucity of tools that can report on the ROS levels within the mitochondria. In particular, there is a lack of tools that can reversibly monitor redox changes over time, providing the potential to sense oxidation-reduction fluxes and to distinguish between transient oxidative bursts and chronic oxidative stress. Although the flavin-based redox probe **NpFR2**, discussed in Chap. 2, demonstrated excellent mitochondrial accumulation, a new generation redox probe that exhibits ratiometric fluorescence properties in addition to mitochondrial localisation would be a valuable addition to the limited ratiometric and reversible mitochondrial redox probes developed to date.

As discussed in Chap. 3, FRET is an elegant strategy to develop ratiometric probes and this was successfully implemented in the case of **FCR1**, a flavin-based ratiometric redox probe with cytoplasmic localisation. This chapter details the work performed towards employing the FRET strategy for the development of a mitochondrially-targeted ratiometric probe with reversible redox sensing abilities. The sub-cellular localisation of the developed probes was interrogated using confo-

Parts of the text and figures of this chapter are reprinted with permission from *Antioxidants and Redox Signalling*, Volume 24, Issue 13, published by Mary Ann Leibert, Inc., New Rochelle, NY.

cal microscopy and subsequent biological experiments were undertaken to evaluate
the reversible redox sensing properties of the probes. Aspects of the work discussed
in this chapter have been published in *Antioxidants and Redox Signalling* [7].

4.1 Excitation-Ratiometric Redox Probes

Considering the inherent redox responsive abilities of flavin, it was again selected
as the redox-active moiety. As outlined in Sect. 3.2, in order to develop a FRET-
based ratiometric probe, it is essential to identify a good donor-acceptor FRET pair,
such as the coumarin-flavin pair in **FCR1**. Mitochondrial localisation can then be
accomplished either by modifying the scaffold to incorporate a mitochondrial tag,
or by employing a different fluorophore with inherent mitochondrial-localising ability (Fig. 4.1), which must also be capable of establishing a good FRET-pair with
flavin. Fluorophores with delocalised cationic nature, such as rhodamine and cyanine derivatives, commonly accumulate in the mitochondria. This accumulation is
dependent on the negative potential across the mitochondrial membrane [8].

Rhodamines are a class of fluorophores that satisfy both these conditions. The
spectral properties of rhodamine indicated that it was ideal to act as a FRET-acceptor
from flavin (Fig. 4.2). With the redox-active component fixed as the FRET donor
and rhodamine as the FRET acceptor, the strategy was to develop an excitation-
ratiometric redox probe. Amongst the FRET-based ratiometric probes reported to
date for a myriad of sensing purposes, emission ratiometric probes dominate the
literature, with only a limited number of probes that are excitation-ratiometric [9,
10]. Development of a redox probe with excitation-ratiometric sensing properties
would be a valuable addition to the list of probes that have employed this strategy.

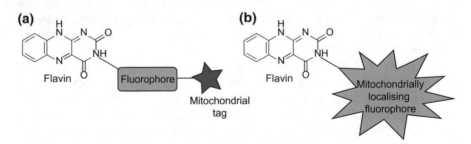

Fig. 4.1 Two approaches towards the design of a mitochondrially-targeted flavin-based ratiometric
redox probe—tethering flavin to a fluorophore **a** that is attached to a mitochondrial tag and **b** with
inherent mitochondrially-localising ability

Fig. 4.2 Absorbance
(dashed line) and emission
(solid line) spectra of flavin
(black) (10 μM) and
rhodamine (red) (10 μM in
100 mM HEPES buffer, pH
7.4) indicating a significant
overlap (yellow) of the
emission profile of the flavin
moiety with the absorbance
of the rhodamine

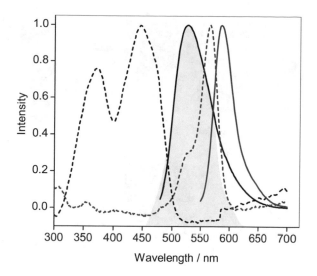

4.2   Flavin-Rhodamine FRET Probe

The rationale behind utilising the excitation-ratiometric flavin-rhodamine FRET pair
as a reporter of redox state is that in the oxidised form, there would be a greater extent
of spectral overlap between the flavin and rhodamine scaffold, and excitation of the
flavin would result in a red rhodamine acceptor emission by FRET. In the reduced
form, owing to the formation of the colourless and non-fluorescent bent conformation
of flavin, the spectral overlap between flavin and rhodamine is minimised, FRET is
suppressed, resulting in negligible emission from the rhodamine acceptor (Fig. 4.3).

Furthermore, excitation at a different (longer) wavelength of the rhodamine itself
for both the oxidised and reduced forms of the flavin would result in red emission,
independent of the FRET process. Therefore, a ratio of the emission intensities of
rhodamine resulting from excitation at two different wavelengths could be used to
gauge the redox state. On the basis of the predicted spectral overlap between flavin
and rhodamine, two flavin-rhodamine redox probes **FRR1** and **FRR2**, were designed
(Fig. 4.4).

Flavin-rhodamine redox probe 1 (**FRR1**) contains N-ethylflavin tethered to rhodamine B *via* piperazine, a relatively rigid spacer. N-Ethylflavin is the redox sensing
group used in **FCR1**. In this study, the use of a naturally-existing flavin derivative,
tetraacetylriboflavin, was also investigated. Tetraacetylriboflavin, with an acetylated-
ribose tail at N-10 and methyl groups at positions 7 and 8 on the isoalloxazine
ring, has been reported to possess a higher quantum yield and significantly red-
shifted excitation and emission wavelengths compared to N-ethylflavin [11]. Therefore, the influence of these structural variations on the photophysical and redox
properties of the probes was investigated. Moreover, it was envisioned that the
lipophilic tetraacetylribose group could potentially improve the cellular retention and

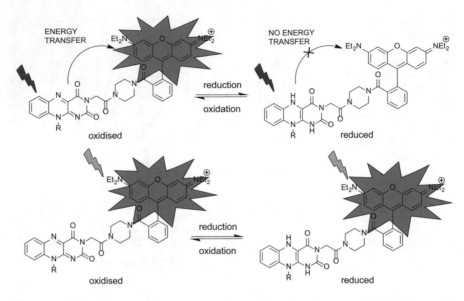

Fig. 4.3 The FRET process taking place within an excitation-ratiometric flavin-rhodamine probe in the oxidised and reduced forms

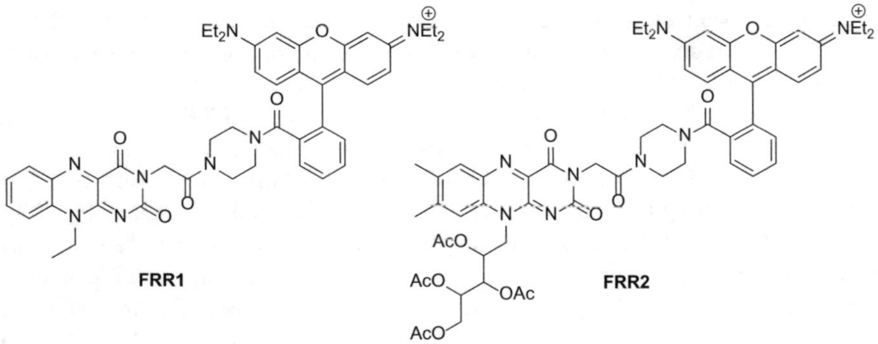

Fig. 4.4 The chemical structures of **FRR1** and **FRR2** Reprinted with permission from *Antioxidants and Redox Signalling*, Volume 24, Issue 13, published by Mary Ann Leibert, Inc., New Rochelle, NY

localisation of the probe. Discussions regarding probe design with Dr. Karolina Jankowska, a postdoctoral researcher in the group, led to the determination of the synthetic approach towards the flavin-rhodamine FRET probes (Scheme 4.1). Synthesis was then carried out by Dr. Jankowska.

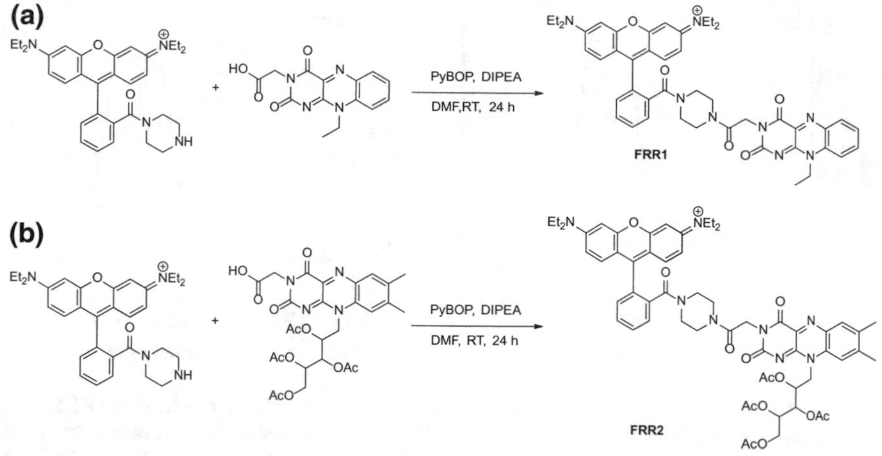

Scheme 4.1 Synthesis of **a FRR1** and **b FRR2** Reprinted with permission from *Antioxidants and Redox Signalling*, Volume 24, Issue 13, published by Mary Ann Leibert, Inc., New Rochelle, NY

4.3 Spectral Characterisation of FRR Probes

Considering the spectral overlap of flavin and rhodamine scaffolds (Fig. 4.2), excitation wavelengths of 460 and 530 nm were chosen to ensure maximum excitation of flavin and rhodamine respectively. Similar to the redox probes investigated thus far, the photophysics of **FRR1** and **FRR2** were studied in HEPES buffer (100 mM, pH 7.4). These initial studies were performed by Dr. Karolina Jankowska using an excitation wavelength of 460 nm because at this wavelength the flavin moiety exhibits high absorbance whereas the rhodamine scaffold has negligible absorbance, thus allowing for preferential excitation of the flavin component over rhodamine in **FRR1** and **FRR2**. These studies indicated that upon excitation at 460 nm, **FRR1** and **FRR2** exhibit two emission maxima: the flavin maximum at 510 nm (**FRR1**) and 525 nm (**FRR2**) respectively; and the rhodamine maximum observed at 580 nm for both the probes (Fig. 4.5). Furthermore, the rhodamine could be independently excited at 530 nm.

FRR1 and **FRR2** were examined for their response to reduction by sodium dithionite. As envisioned, this caused reduction of the flavin molecule to its non-emissive configuration, resulting in a suppressed FRET mechanism. Consequently, upon excitation at 460 nm the integrated emission intensity of both the probes decreased. The ratio of the peak intensity of both **FRR** probes at 580 nm upon excitation at 530 and 460 nm increased upon reduction (Fig. 4.6). Re-oxidation of the reduced probes by treatment with H_2O_2 could be achieved in 20 min, and resulted in the restoration of the original fluorescence properties.

More rigorous characterisation of the redox responsive and spectral characterisation of **FRR1** and **FRR2** were performed by Dr. Jankowska. These studies indicated that **FRR1** exhibited a 3-fold increase in the I_{530ex}/I_{460ex} ratio upon reduction and

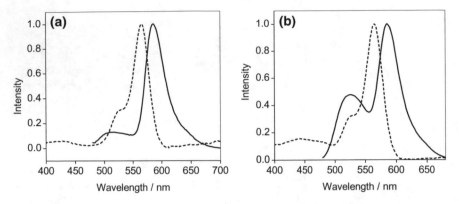

Fig. 4.5 Absorbance (dotted line) and emission (solid line, $\lambda_{ex} = 460$ nm) spectra of (**a**) **FRR1** and (**b**) **FRR2** (10 μM in 100 mM HEPES buffer, pH 7.4). Spectra obtained by Dr. Jankowska Reprinted with permission from *Antioxidants and Redox Signalling*, Volume 24, Issue 13, published by Mary Ann Leibert, Inc., New Rochelle, NY

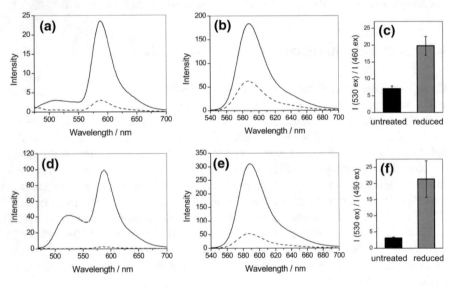

Fig. 4.6 Fluorescence emission spectra of **a, b FRR1** (10 μM) and **d, e FRR2** (10 μM) in oxidised (black) and reduced (dashed line) forms upon excitation at 460 nm (**a, d**) and 530 nm (**b, e**). Probes were reduced using 200 equivalents of $Na_2S_2O_4$. The ratio of the emission of **FRR1** (**c**) and **FRR2** (**f**) at 580 nm upon excitation at 530 versus 460 nm in oxidised (black) and reduced (grey) forms. All data were acquired in 100 mM HEPES buffer, pH 7.4. Error bars represent standard deviation (n = 3). Spectra were obtained by Dr. Jankowska Reprinted with permission from *Antioxidants and Redox Signalling*, Volume 24, Issue 13, published by Mary Ann Leibert, Inc., New Rochelle, NY

the ratiometric response of the probe towards reduction and re-oxidation remained unaltered for up to 5 cycles. **FRR2** displayed a more pronounced 7-fold change in the I_{530ex}/I_{460ex} ratio upon sequential reduction and re-oxidation events, with its ratiometric response remaining unchanged for up to 7 cycles. These results demonstrated the potential of **FRR1** and **FRR2** to function as reversible sensors for redox state. Just as in the case of other redox probes discussed in Chaps. 2 and 3, control experiments were performed on **FRR1** and **FRR2**, which showed that the probes undergo re-oxidation with biological ROS/RNS within 30 min, and the I_{530ex}/I_{460ex} ratio remained unaltered in the presence of biologically relevant metal ions, and at pH values between 3 and 8.

While, rhodamine based probes are widely used for imaging purposes due to their high fluorescence quantum yields, at high concentrations, the fluorescence intensity decreases considerably, and this concentration based quenching is attributed to the formation of dimers [12]. Therefore, to determine the concentrations of the probes that would be ideal for imaging purposes, a concentration-based fluorescence assay was performed by recording the fluorescence spectra of the probes in HEPES buffer (100 mM, pH 7.4) with incremental probe concentrations (Fig. 4.7).

The integrated fluorescence intensities of both **FRR1** and **FRR2** followed a linear trend for concentrations ranging from 5–80 μM. Concentration-based quenching does not occur below 80 μM, therefore probe concentrations below this value should be suitable for biological imaging, provided there are no cytotoxic effects.

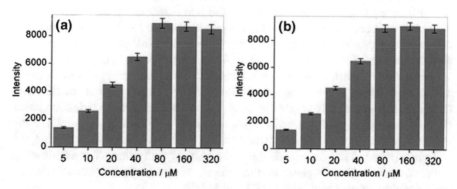

Fig. 4.7 Fluorescence of **FRR1** (**a**) and **FRR2** (**b**) with varying concentration. Bars represent the integrated fluorescence intensity ($\lambda_{ex} = 460$ nm, $\lambda_{em} = 480 - 700$ nm). All data were acquired in 100 mM HEPES buffer, pH 7.4. Errors bars represent standard deviation ($n = 3$) Reprinted with permission from *Antioxidants and Redox Signalling*, Volume 24, Issue 13, published by Mary Ann Leibert, Inc., New Rochelle, NY

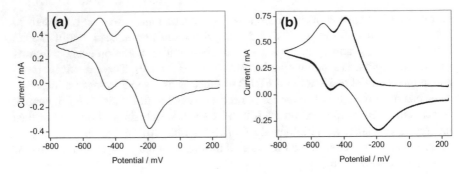

Fig. 4.8 Cyclic voltammogram of **a** FRR1 and **b** FRR2 (5 mM concentration) in MeCN at 25°C with a scan rate of 100 mV/s Reprinted with permission from *Antioxidants and Redox Signalling*, Volume 24, Issue 13, published by Mary Ann Leibert, Inc., New Rochelle, NY

4.4 Electrochemistry

The reduction potentials of **FRR1** and **FRR2** were investigated by recording cyclic voltammograms with 5 mM concentrations of the probes in MeCN containing TBAB (tetrabutylammonium bromide) as an electrolyte. The cyclic voltammograms depicted two sets of peaks for each probe (Fig. 4.8). The peaks at the lower half-wave potential (E°) of −544 mV correlates to the value reported for the reduction potential of rhodamine B [13]. The peaks at −186 and −369 mV (E° = −259 mV vs. SHE) in **FRR1** correspond to the reduction potential of the *N*-ethylflavin component, whereas the peaks at −173 and −413 mv (E° = −290 mV vs. SHE) observed for **FRR2** represents the reduction potential of tetraacetylriboflavin (Fig. 4.9)

These electrochemical studies demonstrate that varying the substituent at the *N*-10 position on the isoalloxazine ring modulates the reduction potential of the flavin. In this case, it was observed that the ribose tail in **FRR2** tuned the reduction potential to a more biologically relevant value compared to **FRR1**. Although the cyclic voltammograms indicate a peak corresponding to the reduction of rhodamine, this reduction potential lies far outside the scope of biological values. The profile of the obtained cyclic voltammograms confirm the chemical reversibility of the oxidation and reduction processes of both **FRR1** and **FRR2**.

4.4.1 Fluorescence Properties at 488 nm Excitation

While the ratiometric response of **FRR1** and **FRR2** towards redox state has been extensively investigated, the use of these probes depends upon the availability of two appropriate excitation sources, which many standard microscopy and cytometry instruments lack. In order to illustrate the broader applicability of both probes in such protocols, the redox-responsive ability of the probes under a single excitation wavelength was investigated. Upon excitation at 488 nm, **FRR1** did not exhibit any fluorescence emission from the flavin and only minimal fluorescence from the rho-

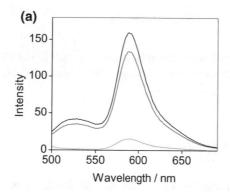

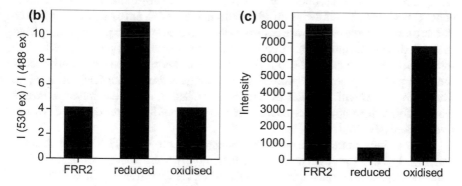

Fig. 4.9 Photophysical behaviour of **FRR2** under 488 nm excitation. **a** Spectra of **FRR2** untreated (black), reduced with 2 mM sodium dithionite (red) and re-oxidised with 4 mM H_2O_2 (blue). **b** Fluorescence emission ratio of emission intensity upon excitation with 530 and 488 nm. **c** Integrated intensity of red fluorescence ($\lambda_{ex} = 488$ nm, $\lambda_{em} = 560 - 700$ nm) All spectra were acquired at 10 μM concentration of probe in 100 mM HEPES buffer, pH 7.4 Reprinted with permission from *Antioxidants and Redox Signalling*, Volume 24, Issue 13, published by Mary Ann Leibert, Inc., New Rochelle, NY

damine. Therefore, upon excitation at 488 nm no significant changes were observed in the fluorescence properties of **FRR1** upon reduction.

Although in the case of **FRR2** (Fig. 4.9a) the peak corresponding to the fluorescence from flavin was greatly diminished, a similar trend in ratiometric response to reduction was observed upon excitation at 488 nm (I_{530ex}/I_{488ex}) as for 460 nm (I_{530ex}/I_{460ex}). The decrease in the (I_{530ex}/I_{488ex}) ratio upon reduction could be reversed upon re-oxidation with H_2O_2 (Fig. 4.9b). Notably, the absolute fluorescence intensity (560–590 nm) of **FRR2** decreases upon reduction, and this decrease can be reversed by treatment with H_2O_2 (Fig. 4.9c). Therefore, **FRR2** demonstrates the ability to report on variations in the redox environment, when excited using a single wavelength excitation of 488 nm. Hereafter, microscopy and flow cytometry experiments were performed using a 488 nm excitation wavelength, with a more oxidised probe indicated by a more intense red fluorescence.

4.5   Biological Imaging Experiments

Having established the redox-active behaviour and reversibility of **FRR1** and **FRR2**, the probes were investigated for their cellular localisation, toxicity and ability to report on biological oxidative capacity.

4.5.1   Sub-cellular Localisation

The sub-cellular localisation of the probes was examined in RAW 264.7 macrophages. Using a confocal microscope, cells were excited by a 488 nm laser and the fluorescence emission was recorded from 495–620 nm. Under 488 nm excitation, negligible fluorescence was observed from control cells untreated with the probe. As depicted in Fig. 4.10, significantly higher fluorescence intensity was observed when cells were treated with **FRR1** or **FRR2** (20 μM, 15 min). Furthermore, the sub-cellular accumulation pattern of both the probes strongly suggested mitochondrial localisation. Therefore, just as with **NpFR2**, as discussed in Chap. 2, colocalisation experiments were performed employing commercial tracker dyes—Mitotracker DeepRed FM and Lysotracker DeepRed.

Singly stained controls, prepared by treating cells with **FRR1**, **FRR2**, Mitotracker DeepRed FM and Lysotracker DeepRed individually, were imaged to ensure that no emission leaked from one channel into the other. As shown in Fig. 4.10, significant fluorescence was observed from cells treated with **FRR1** or **FRR2** (20 μM, 15 min) in channel 1 (λ_{ex} 488 nm, λ_{em} 495 − 620 nm), whilst negligible fluorescence observed in channel 2 (λ_{ex} 633 nm, λ_{em} 650 − 750 nm). Cells treated with the tracker dyes fluoresced only in channel 2, thus validating that there is no fluorescence bleed-through from one channel to another.

Cells were then co-stained by treating them with **FRR1** or **FRR2** (20 μM, 15 min) and Mitotracker DeepRed FM (100 nM, 15 min). The cells were interrogated for their fluorescence properties in channels 1 and 2, and pseudo-coloured green and red respectively. The fluorescence images from both the channels were then merged using FIJI (National Institutes of Health), an image processing software. As indicated by the yellow regions in the merged image in Fig. 4.11, fluorescence emission of both **FRR1** and **FRR2** overlaps significantly with the commercial Mitotracker DeepRed FM, confirming clear mitochondrial accumulation of the probes. RAW 264.7 cells co-stained with the probes and the Lysotracker DeepRed (100 nM, 15 min), illustrate very different localisation profiles (Fig. 4.11), suggesting very low association with lysosomes. This was further evidenced by the poor Pearson's co-localisation coefficients calculated to be 0.25 and 0.13 for **FRR1** and **FRR2**, respectively.

The mitochondrial localisation of **FRR1** and **FRR2** is further evidenced by the plot profiles (Fig. 4.12), in which a line selection is made across the image and the intensities of fluorescence in each of the channels are then plotted against the distance. A complete overlap of the grey values (fluorescence intensity) from both

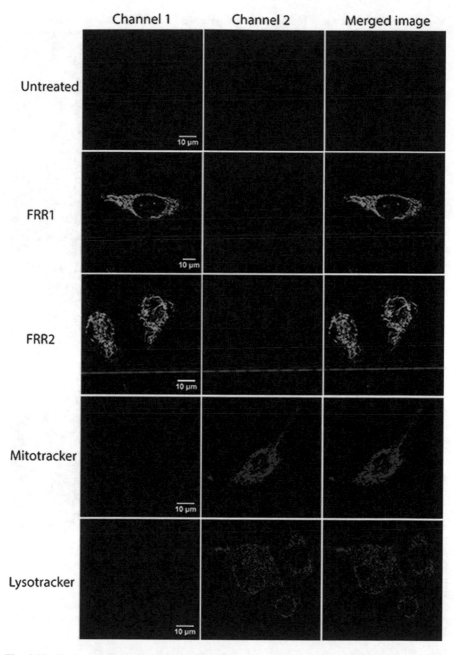

Fig. 4.10 Confocal microscopy images of RAW 264.7 cells, untreated, cells treated only with **FRR1** (20 μM, 15 min), **FRR2** (20 μM, 15 min), Mitotracker Deep Red (100 nM, 15 min) and Lysotracker Deep Red (100 nM, 15 min), in channel 1 ($\lambda_{ex} = 488$ nm, $\lambda_{em} = 495 - 600$ nm), channel 2 (100 nM, $\lambda_{ex} = 633$ nm, $\lambda_{em} = 650 - 750$ nm) and merged images of channel 1 and 2 Reprinted with permission from *Antioxidants and Redox Signalling*, Volume 24, Issue 13, published by Mary Ann Leibert, Inc., New Rochelle, NY

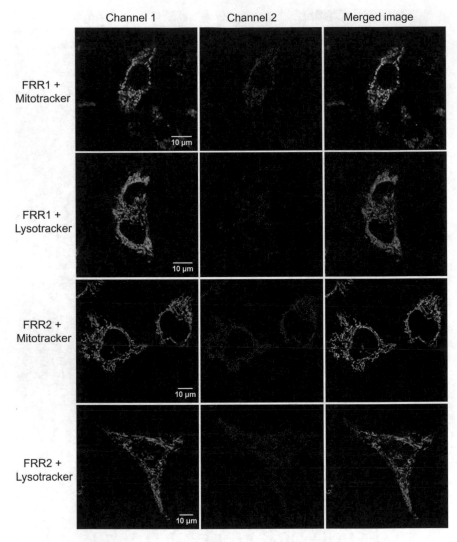

Fig. 4.11 Co-localization images of macrophages (RAW 264.7) treated with **FRR1** (20 μM) or **FRR2** (20 μM), co-stained with Mitotracker deep red (100 nM) and Lysotracker deep red (100 nM). **FRR1/FRR2** emission is in channel 1 ($\lambda_{ex} = 488$ nm, $\lambda_{em} = 495 - 620$ nm) and Mitotracker/Lysotracker emission in channel 2 ($\lambda_{ex} = 633$ nm, $\lambda_{em} = 650 - 750$ nm). Merged images indicate good co-localisation of Mitotracker with both probes Reprinted with permission from *Antioxidants and Redox Signalling*, Volume 24, Issue 13, published by Mary Ann Leibert, Inc., New Rochelle, NY

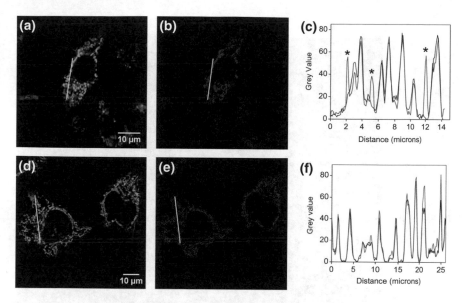

Fig. 4.12 Plot profiles of **FRR1** (**a**) and **FRR2** (**d**) in comparison to Mitotracker Deep Red (**b**, **e**). The black lines in the plot profiles (**c**, **f**) represent Mitotracker Deep Red and those for FRR1 (**c**) and FRR2 (**f**) are shown in red. Profiles have been generated from the regions marked with a white line in the images. ∗indicates regions where there is **FRR1** fluorescence but no MitoTracker emission Reprinted with permission from *Antioxidants and Redox Signalling*, Volume 24, Issue 13, published by Mary Ann Leibert, Inc., New Rochelle, NY

the channels would indicate the presence of both probe and the tracker dyes at each pixel, suggesting a perfect colocalisation. The plot profile of **FRR1** indicated the presence of some peaks (marked with an asterisk, Fig. 4.12c) that did not overlap with the Mitotracker. This indicates that there are some non-mitochondrial regions in the cell where **FRR1** localises. However, the plot profile of **FRR2** exhibits great extent of overlap with that of Mitotracker (Fig. 4.12). The Pearson's co-localisation coefficients were determined to be 0.68 and 0.92 for **FRR1** and **FRR2** respectively which is in agreement with the results obtained from the plot profiles for both the probes.

In the interests of further validating the mitochondrial localisation of the probes, co-localisation experiments were carried out with the commercially available CellLight® Mitochondria-GFP, BacMam 2.0. This is a cell-transfection based technology in which mithochondrial targeting is achieved by E1 alpha pyruvate dehydrogenase (a mitochondrial enzyme) and visualisation by a fluorescent protein. A fused DNA construct for E1 alpha pyruvate dehydrogenase and emerald green fluorescent protein (emGFP) is packaged within a Baculovirus and transfection of cells with these virus particles result in the production of a fluorescent mitochondrial enzyme. This enables the detection of mitochondria independent of mitochondrial membrane potential. Since this technology cannot be applied to macrophages, DLD-1 human colorectal cancer carcinoma cells were investigated. DLD-1 cells were transfected

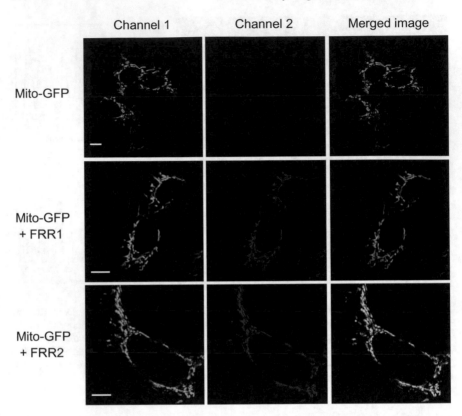

Fig. 4.13 Representative confocal microscopy images of DLD-1 cells, cells transfected with CellLight® mitochondria-GFP, BacMam 2.0, co-stained with **FRR1** (20 μM, 15 min), **FRR2** (20 μM, 15 min), in channel 1 ($\lambda_{ex} = 488$ nm, $\lambda_{em} = 510 - 540$ nm), channel 2 ($\lambda_{ex} = 488$ nm, $\lambda_{em} = 560 - 700$ nm) and merged images of channel 1 and 2. Scale bars represent 10 μm Reprinted with permission from *Antioxidants and Redox Signalling*, Volume 24, Issue 13, published by Mary Ann Leibert, Inc., New Rochelle, NY

with CellLight® Mitochondria-GFP, BacMam 2.0 overnight at a concentration of 10 particles per cell (PPC) followed by treatment with **FRR1** or **FRR2** (20 μM, 15 min). Excellent mitochondrial colocalisation was observed for both **FRR1** and **FRR2**, as indicated by the yellow regions in the merged image (Fig. 4.13).

To ensure that the probes did not have any toxic effects, cytotoxicity of the probes was evaluated using the standard MTT assay. This was accomplished by treating RAW 264.7 murine macrophage cells with **FRR1** and **FRR2** at 0–160 μM concentrations and incubated for a period of 24 h followed by 4 h of treatment with MTT. The IC_{50} values were determined to be 38 (± 1 μM) and 41 (± 2 μM) for **FRR1** and **FRR2** respectively. These values are much higher both in terms of concentration and treatment times employed in cellular imaging studies. In addition, time-lapse imaging experiments were carried out to examine the consequences of longer laser exposure on the photophysical behaviour of the probe and cell viability. Following

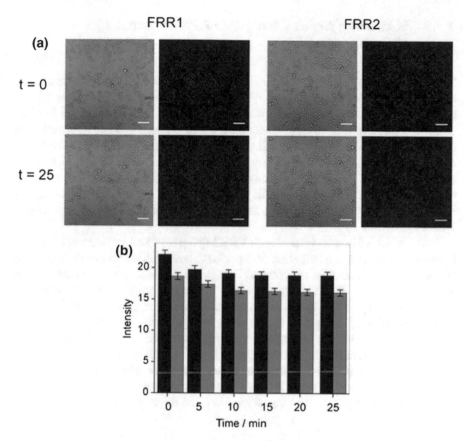

Fig. 4.14 Cell morphology and fluorescence intensity of RAW 264.7 macrophages treated with **FRR1** or **FRR2** (20 μM) and imaged every 30 s with laser excitation at 488 nm, and an acquisition speed of 400 Hz. **a** Transmitted light and confocal ($\lambda_{ex} = 488$ nm, $\lambda_{em} = 560 - 700$ nm) images of cells at 0 and 25 min time-points. **b** Emission intensities obtained at indicated time-points for cells treated with **FRR1** (grey) or **FRR2** (black). Bars represent average fluorescence intensity ($\lambda_{ex} = 488$ nm, $\lambda_{em} = 560 - 700$ nm), errors bars represent standard deviation (n = 3) Reprinted with permission from *Antioxidants and Redox Signalling*, Volume 24, Issue 13, published by Mary Ann Leibert, Inc., New Rochelle, NY

laser irradiation at 488 nm every 30 s, cell morphology and fluorescence emission from RAW 264.7 cells treated with **FRR1** or **FRR2** (20 μM, 15 min) were monitored. The results indicated that **FRR1** and **FRR2** did not exhibit any self-amplication of the fluorescence signal, over 25 min (Fig. 4.14).

Moreover, no significant changes in cell morphology were observed over this time, demonstrating a lack of phototoxic effects. Consistent with the the brighter-green fluorescence from the riboflavin moiety in **FRR2** compared to the synthetic *N*-ethylflavin in **FRR1** aided in confocal imaging and image analysis. Therefore, all further biological experiments were carried out with **FRR2**.

4.5.2   Measuring Redox Changes in LPS-Stimulated Macrophages

The reversibility of redox responsive **FRR2** was harnessed to monitor changes in oxidative capacity of macrophages following a lipopolysaccharide (LPS) insult. LPS, a major outer membrane component of gram negative bacteria, binds to toll-like receptor 4 (TLR4), present on the macrophage membrane, responsible for pathogen recognition [14]. This binding stimulates a cascade of events within the macrophage cell which execute the bactericidal function. An important participant in these events are the mitochondria within the macrophages, which are known to produce a burst of ROS/RNS which subsequently kills the bacteria [14]. Therefore, this was considered an interesting biological system to test the reversible redox-active properties of **FRR2**.

RAW 264.7 cells were treated with **FRR2** (20 μM, 15 min) followed by stimulation with LPS at a concentration of 1 μg/mL, over treatment times ranging from 0–2 h. The cells were then washed and resuspended in FACS buffer (PBS supplemented with 1% FCS) containing propidium iodide (1 μg/mL). The cells were then interrogated for their red fluorescence emission using the 585/42 nm (42 nm band width, centred at 585 nm) filter flow cytometer equipped with a 488 nm laser. For each sample, a population of single and live cells was gated out and the red fluorescence intensity was analysed. As shown in Fig. 4.15, a 5-fold increase in fluorescence intensity was observed by the 30 min treatment timepoint, after which the intensity decreases, demonstrating the burst in ROS production takes place 30 min after the LPS stimulation. By 2 h after LPS stimulation, the fluorescence emission returned

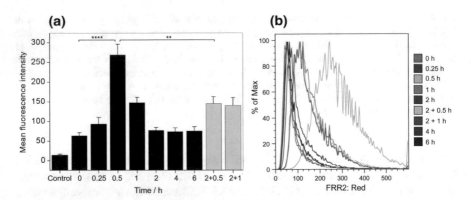

Fig. 4.15   a Response of **FRR2** to macrophages (RAW 264.7) stimulated with LPS from 0–6 h (black) and those re-stimulated with LPS after the 2 h time-point (grey). Bars represent the mean fluorescence intensity of red emission (585/42 nm) when excited with a 488 nm laser. Error bars represent standard error of mean, $p < 0.05$ is considered significant. **b** Representative flow cytometry dataset. The histograms represent intensity of red fluorescence emissions (585/42 nm) when excited with a 488 nm laser. Reprinted with permission from *Antioxidants and Redox Signalling*, Volume 24, Issue 13, published by Mary Ann Leibert, Inc., New Rochelle, NY

to levels observed for unstimulated cells. Similar emission levels were observed at 4 and 6 h after LPS treatment.

Following this, a re-stimulation experiment was performed, wherein the macrophages were re-stimulated with LPS 2 h after the initial stimulation. Although there was a significant increase in fluorescence intensity after 0.5 and 1 h, this increase was less drastic than the initial response. This experiment demonstrate the ability of **FRR2** to report on the changes in the mitochondrial oxidative capacity of RAW 264.7 cells following an LPS-insult, clearly illustrate the potential of **FRR2** as a reversible sensor of redox fluxes within the mitochondria.

This result was further confirmed by live cell imaging experiments. RAW 264.7 macrophages were stimulated with LPS for 0.5 and 1 h followed by treatment with **FRR2** (Fig. 4.16). Similar to the results obtained from flow cytometry, a significant increase in intensity was observed 30 min post-stimulation, followed by a decrease thereafter, validating that a maximal burst in the mitochondrial ROS production

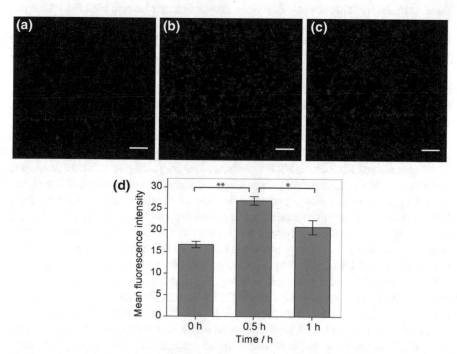

Fig. 4.16 Confocal microscope images of RAW 264.7 macrophages cells showing the fluorescence intensity ($\lambda_{ex} = 488$ nm, $\lambda_{em} = 560 - 700$ nm) upon stimulation with LPS for 0 h (**a**), 0.5 h (**b**) and 1 h (**c**) followed by treatment with **FRR2** (20 μM, 15 min). Scale bars represent 100 μm. **d** Bars represent the mean fluorescence intensity of images acquired from macrophages stimulated with LPS for indicated times. Error bars represent standard deviation, $^*p < 0.05$, $^{**}p < 0.01$ Reprinted with permission from *Antioxidants and Redox Signalling*, Volume 24, Issue 13, published by Mary Ann Leibert, Inc., New Rochelle, NY

occurs 30 min after the LPS-insult. The obtained results establish the potential of **FRR2** to reversibly report on the variations in mitochondrial redox state.

4.6   Conclusions

Although excitation-ratiometric probes are less commonly designed and developed because of the complications that arise from the simultaneous use of two different excitation wavelengths, they are certainly more beneficial than intensity-based probes. With their excitation-ratiometric output, **FRR1** and **FRR2** show great potential as fluorescent tools to interrogate fluxes in oxidative capacity within the mitochondria. Mitochondrial targeting of FRET-based probes can be achieved by taking advantage of fluorophores with intrinsic mitochondrial localising abilities. This removes the requirement of attaching a mitochondrial targeting tag, mitigating synthetic complexity. This chapter describes the studies that investigated the properties of two flavin-rhodamine redox probes **FRR1** and **FRR2**, which differ by the substitution at the N-10 position on the flavin scaffold. **FRR1** contains a synthetic N-ethylflavin, whilst **FRR2** incorporates a naturally existing riboflavin. The electrochemical studies confirmed that the ribose group at the N-10 position on the flavin scaffold of **FRR2** was capable of modulating the reduction potential of the probe to a more biologically-relevant range—this aspect of probe design could be applicable towards the development of redox probes with different redox-active potentials.

In addition, co-localisation experiments indicated clear mitochondrial localisation of the probes **FRR1** and **FRR2**. The fluorescence properties indicated a brighter-flavin emission of **FRR2** than **FRR1**, aiding its imaging using a confocal microscope and image analysis thereafter. Furthermore, it is important to ensure that the photophysical properties of a developed probe are not too exotic to limit its application to limited state of the art instrumental set-ups. The broader applicability of probes for much simpler protocols may also prove beneficial. Therefore, the suitability of the probe to measure redox changes using a basic flow cytometer was examined. The results from the LPS stimulation experiment validate it to be a valuable system for the characterisation of a reversible redox probe, particularly for the mitochondria, such as **FRR2**. The probe demonstrated that a burst in the production of mitochondrial ROS in macrophages occurs following LPS stimulation and that this burst is the greatest 30 min post LPS stimulation. In addition, more rigorous investigations using **FRR2** to report on the variations in mitochondrial oxidative capacity in specific biological systems are discussed in Chaps. 6 and 7.

References

1. P.H.G.M. Willems, R. Rossignol, C.E.J. Dieteren, M.P. Murphy, W.J.H. Koopman, Redox homeostasis and mitochondrial dynamics. Cell Metab. **22**, 207–18 (2015)

2. M.P. Murphy, A. Holmgren, N.-G. Larsson, B. Halliwell, C.J. Chang, B. Kalyanaraman, S.G. Rhee, P.J. Thornalley, L. Partridge, D. Gems, T. Nystrom, V. Belousov, P.T. Schumacker, C.C. Winterbourn, Unraveling the biological roles of reactive oxygen species. Cell Metab. **13**, 361–366 (2011)
3. C.L. Bigarella, R. Liang, S. Ghaffari, Stem cells and the impact of ROS signaling. Development **141**, 4206–4218 (2014). (Cambridge, England)
4. C. Wolke, A. Bukowska, A. Goette, U. Lendeckel, Redox control of cardiac remodeling in atrial fibrillation. Biochim. Biophys. Acta **2015**, 1555–1565 (1850)
5. A.K. Biala, R. Dhingra, L.A. Kirshenbaum, Mitochondrial dynamics: orchestrating the journey to advanced age. J. Mol. Cell. Cardiol. **83**, 37–43 (2015)
6. L.F.R. Francesca, *Bonomini Rita Rezzani, Metabolic Syndrome, Aging and Involvement of Oxidative Stress*
7. A. Kaur, K. Jankowska, C. Pilgrim, S.T. Fraser, E.J. New, *Studies of hematopoietic cell differentiation with a ratiometric and reversible sensor of mitochondrial reactive oxygen species* (Antioxid, Redox Signal, 2016)
8. R.A. Smith, R.C. Hartley, M.P. Murphy, Mitochondria-targeted small molecule therapeutics and probes. Antioxid. Redox Signal. **15**, 3021–3038 (2011)
9. L. Yuan, W. Lin, Z. Cao, J. Wang, B. Chen, Development of FRET-based dual-excitation ratiometric fluorescent pH probes and their photocaged derivatives. Chemistry (Weinheim an der Bergstrasse, Germany) **18**, 1247–1255 (2012)
10. L. Yuan, Q.-P. Zuo, FRET-based mitochondria-targetable dual-excitation ratiometric fluorescent probe for monitoring hydrogen sulfide in living cells. Chem.-An Asian J. **9**, 1544–1549 (2014)
11. M. Insińska-Rak, A. Golczak, M. Sikorski, Photochemistry of riboflavin derivatives in methanolic solutions. J. Phys. Chem. A **116**, 1199–1207 (2012)
12. D. Setiawan, A. Kazaryan, M.A. Martoprawiro, M. Filatov, A first principles study of fluorescence quenching in rhodamine B dimers: how can quenching occur in dimeric species? Phys. Chem. Chem. Phys. **12**, 11238–11244 (2010)
13. T. Takizawa, T. Watanabe, K. Honda, Photocatalysis through excitation of adsorbates. 2. A comparative study of Rhodamine B and methylene blue on cadmium sulfide. J. Phys. Chem. **82**, 1391–1396 (1978)
14. A.P. West, I.E. Brodsky, C. Rahner, D.K. Woo, H. Erdjument-Bromage, P. Tempst, M.C. Walsh, Y. Choi, G.S. Shadel, S. Ghosh, TLR signalling augments macrophage bactericidal activity through mitochondrial ROS. Nature **472**, 476–480 (2011)

Chapter 5
Nicotinamide Based Ratiometric Redox Probes

As detailed in Chap. 1, a myriad of metabolic processes occur within a biological system, including many redox reactions, each operating at its own redox potential. In order to obtain a deeper understanding of cellular oxidative capacity, it is important to gain a wholistic picture of cellular redox processes and the extent to which each process contributes towards the cell's redox homoeostasis. It is therefore necessary to develop a suite of redox sensors, each with its own redox potential, that can be used together to accurately image a range of biological redox environments. While the dynamic range of flavin-based probes, discussed in Chaps. 2 and 3, spans the redox potentials of a variety of key redox processes, the aim of this section of the work was to identify another scaffold active over different redox potentials. One such redox-responsive molecule within a cell is nicotinamide (vitamin B_3). This chapter discusses the design and synthesis of fluorescent redox probes based on nicotinamide as the redox responsive component, along with the characterisation and biological studies of the developed probes.

5.1 Nicotinamide as a Redox Responsive Scaffold

Structurally incorporated within two vital reducing currencies of the cell—nicotinamide adenine dinucleotide (NADH) and nicotinamide adenine dinucleotide phosphate (NADPH), nicotinamide acts as the redox-sensitive component of these coenzymes, catalysing electron transfer to and from different metabolic processes (Fig. 5.1) [1].

The redox-responsive properties of nicotinamide emanate from the ability of the amide-bearing pyridine ring to donate and accept electrons, and cycle between the reduced (NADH or NADPH) and oxidised (NAD$^+$ or NADP$^+$) forms (Fig. 5.2). The redox potential of the NAD$^+$/NADH redox couple is -316 mV versus SHE, which

© Springer International Publishing AG 2018
A. Kaur, *Fluorescent Tools for Imaging Oxidative Stress in Biology*,
Springer Theses, https://doi.org/10.1007/978-3-319-73405-7_5

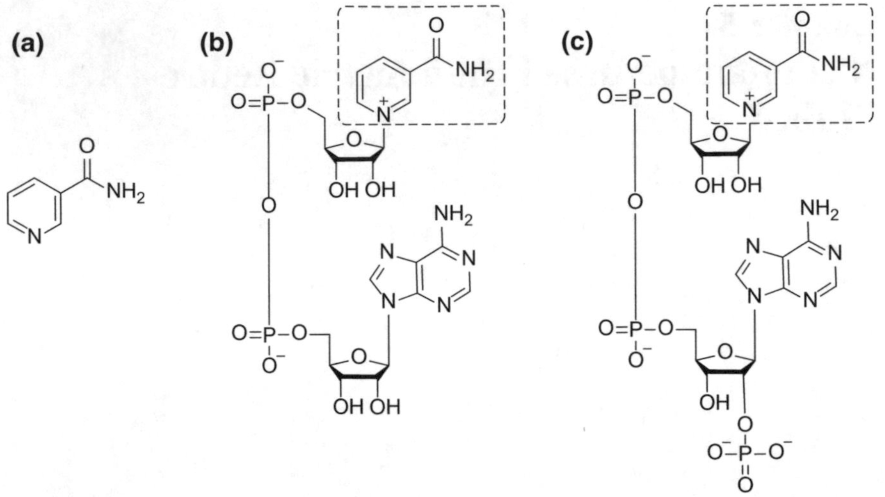

(a)   **(b)**   **(c)**

Fig. 5.1 The chemical structure of **a** nicotinamide, and the oxidised forms of **b** nicotinamide adenine dinucleotide (NADH) and **c** nicotinamide adenine dinucleotide phosphate (NADPH)

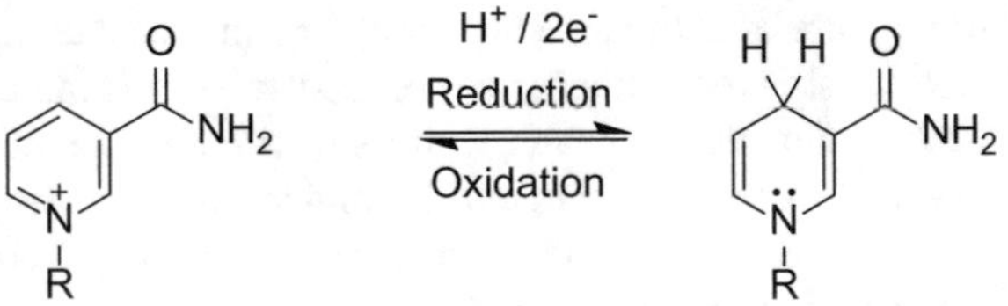

Fig. 5.2 The oxidised and reduced forms of the nicotinamide scaffold

is much lower than that of flavin ($-200\,$mV versus SHE) [2, 3]. This means that a redox sensor based on a nicotinamide scaffold could be employed to interrogate redox processes that take place at more negative potentials.

5.2 Designing a Ratiometric Sensor

The oxidised and reduced forms of nicotinamide have very distinct structural and electron distribution patterns. While a quarternary pyridinium cation is a characteristic feature of the oxidised form, reduction results in the addition of two electrons to the ring, with restoration of the lone pair of electrons on the nitrogen atom (Fig. 5.2). These interesting differences suggest that the oxidised and reduced forms could differ considerably in their ability to make electrons available to a conjugated fluorophore. Although nicotinamide does not have inherent fluorescent properties, the drastic increase in the nucleophilicity of reduced nicotinamide can be exploited to modulate the intensity and/or wavelength of an attached fluorophore.

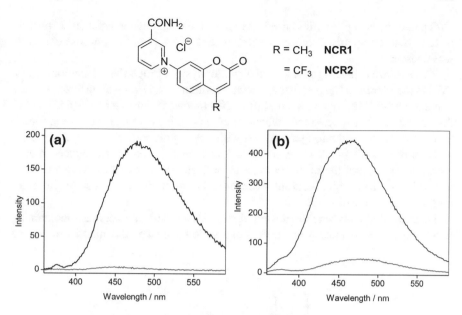

Fig. 5.3 The chemical structures of **NCR1** and **NCR2** (top) and the fluorescence spectra (bottom) of oxidised (black) and reduced (red) forms of **a NCR1** and **b NCR2**, $\lambda_{ex} = 330$ nm

Previous work in the group involved attaching fluorophores such as 7-amino-4-methylcoumarin (coumarin 120) and 7-amino-4-trifluoromethylcoumarin (coumarin 151) directly to the pyridium nitrogen to give nicotinamide coumarin redox sensors 1 and 2 (**NCR1** and **NCR2**, Fig. 5.3).

The redox-sensing mechanism of these probes relied on the photoinduced electron transfer mechanism. Upon reduction the electron-rich nicotinamide scaffold exhibits enhanced PET quenching ability, consequently decreasing the fluorescence intensity of the coumarin moiety. Both **NCR1** and **NCR2** exhibited a decrease in fluorescence intensity upon reduction (Fig. 5.3a, b), but re-oxidation could not be achieved upon treatment of the reduced probes with H_2O_2. **NCR2** has stronger electron-withdrawing properties due to the presence of trifluoromethyl group at position 4, therefore the photophysical properties of **NCR2** were red-shifted compared to **NCR1**. However, the excitation maximum still lay in the UV region making it difficult to visualise these probes in cellular imaging experiments. Therefore, the aim of this section of the work was to investigate alternative structural modifications on the nicotinamide scaffold in order to develop a probe that exhibits significantly longer excitation-emission profiles. Furthermore, attention was focussed towards achieving a ratiometric fluorescence output.

In order to achieve these aims, an alternative position of attachment of the fluorophore to the nicotinamide scaffold was considered, which involved conjugating the fluorophore *via* a carbon atom on the nicotinamide heterocycle instead of the N-atom. In addition, the probe design included an inherently ratiometric fluorophore with photophysical properties based on intramolecular charge transfer (ICT) process to achieve a ratiometric fluorescence response (Fig. 5.4). For this purpose coumarin was chosen.

The extensive investigations of the photophysical properties of coumarin that exist in the literature suggest that an electron withdrawing group at position 3 on the coumarin scaffold results in a red-shift in the coumarin's fluorescence (Fig. 5.4) [4]. As a result, it was expected that, when tethered *via* this position, the oxidised nicotinamide would exhibit stronger electron withdrawing properties and cause a greater bathochromic shift in the coumarin fluorescence compared to the reduced nicotinamides. This could therefore serve as an indicator of redox state. As represented in Fig. 5.5, probe design therefore incorporated a coumarin scaffold conjugated to a nicotinamide at position 3.

One essential feature in the design of a nicotinamide—and coumarin-containing redox sensor is to ensure extended conjugation between the nicotinamide and

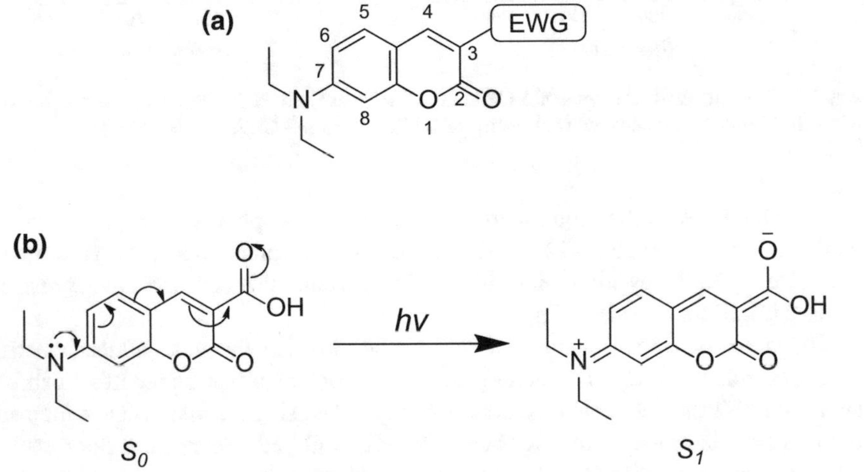

Fig. 5.4 **a** The coumarin scaffold, indicating the preferred position for incorporating an electron withdrawing group and **b** a representation of the intramolecular charge transfer process in a coumarin derivative (7-diethylaminocoumarin-3-carboxylic acid). The diethylamino group acts as the electron donor, while the oxygen of the carboxylic group acts as the acceptor, giving the excited state (S_1)

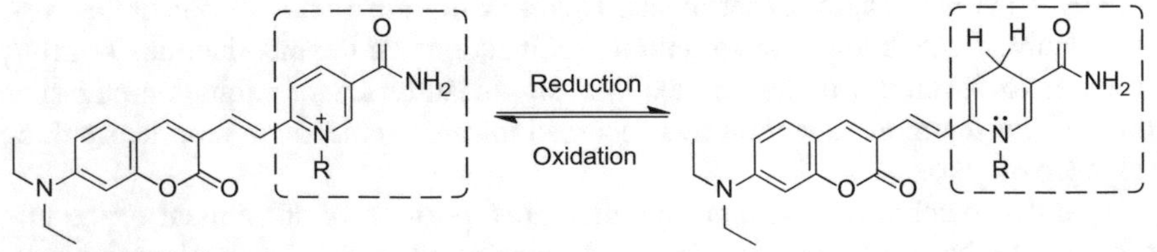

Fig. 5.5 The structures of the oxidised and reduced forms of a rationally designed probe that employs nicotinamide as a redox sensing scaffold (dashed box) and a coumarin fluorophore

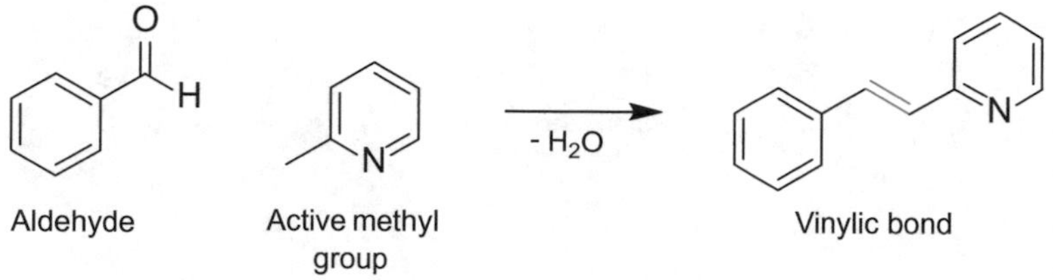

Fig. 5.6 An example of a Knoevenagel-type condensation reaction between an aromatic aldehyde and active methyl group resulting in the formation of a vinylic bond (red) between the two scaffolds

coumarin scaffolds in the final chemical structure. For this purpose, Knoevenagel-type condensation was employed, in which two molecules can be tethered by creating a vinylic bond between them (Fig. 5.6). In this condensation, aromatic aldehydes are reacted with activated methyl groups, such as on aromatic rings bearing electron withdrawing groups.

To achieve this transformation, it was decided to use a coumarin bearing an aldehyde group at position 3 and a nicotinamide bearing a methyl group. Therefore, reaction of a substituted coumarin-3-aldehyde with a methyl nicotinamide was pursued. Furthermore, owing to its well-suited photophysical properties for biological imaging, 7-diethylaminocoumarin was utilised in the synthesis.

5.2.1 Synthesis of Redox Sensors Based on Nicotinamide and Coumarin

The ethyl ester of 7-diethylaminocoumarin-3-carboxylic acid (**46**) was synthesised by the Knoevenagel condensation of 4-diethylaminosalicylaldehyde (**39**) and diethylmalonate in the presence of pyridine as a base (Scheme 5.1). Decarboxylation of **46** was achieved by heating under reflux conditions in acetic acid and concentrated hydrochloric acid, resulting in the formation of 7-diethylaminocoumarin (**47**). A Vilsmeier-Haack reaction was then performed using DMF and phosphorous oxychloride (POCl$_3$), to convert **47** into 7-diethylaminocoumarin-3-aldehyde (**48**).

To understand the effect of the position of methyl group on the ICT mechanism of the final probe, commercially available 5-methylnicotinic acid (**49a**) and 6-methylnicotinic acid (**49b**) were esterified using methanol and concentrated sulfuric acid, giving (**50a, b**), which were then converted into corresponding amides (**51a, b**) upon reaction with a 28% ammonium hydroxide solution (Scheme 5.4).

Knoevenagel-type condensation reactions of the aldehyde **48** with the methyl group in 5- and 6-methyl nicotinamides (**51a, 51b**) were performed in anhydrous DMF in the presence of p-toluenesulfonic acid under N$_2$ atmosphere at RT. The condensation reaction was successful, with condensation of **48** and **51b** resulting in the formation of a red fluorescent nicotinamide-coumarin conjugate (**NCC**) with an

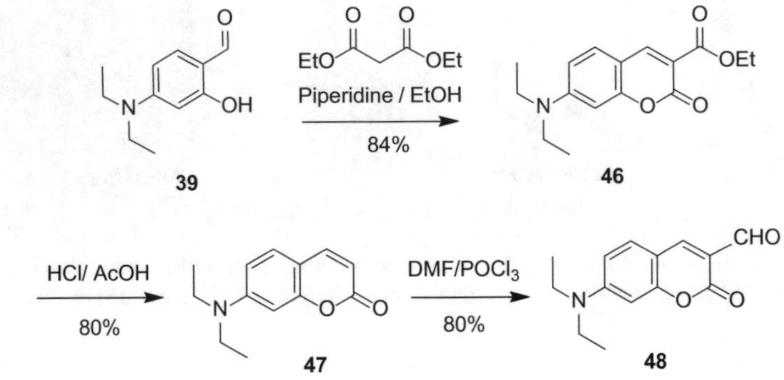

Scheme 5.1 Synthesis of 7-diethylaminocoumarin-3-aldehyde

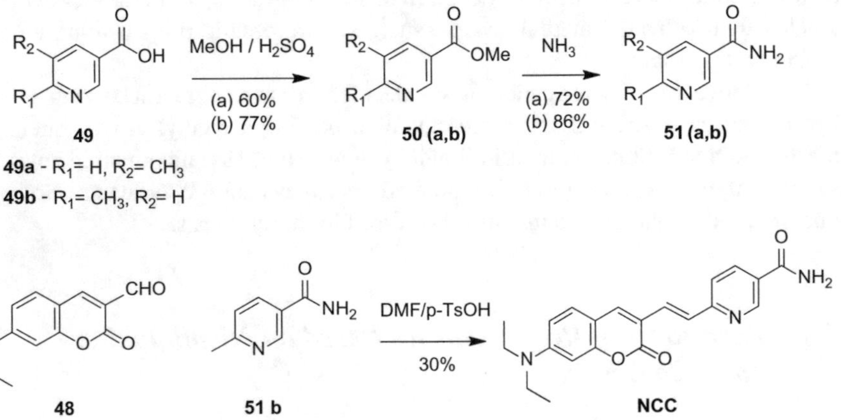

Scheme 5.2 Synthetic scheme for the formation of 5- and 6-methyl nicotinamides and the condensation of 6-methylnicotinamide with the coumarin aldehyde

overall yield of 30%. However, no reaction progress was observed when **51a** was used, possibly because of a deactivated methyl group at position 5 on the ring.

5.2.2   Structural Modifications to Achieve Redox Activity

Following its purification by column chromatography, ^{1}H NMR-spectrum of **NCC** exhibited a doublet at $\delta = 8.15$ ppm, thus confirming the presence of the oxidised form of nicotinamide. Photophysical characterisation of **NCC**, performed in HEPES buffer (100 mM, pH 7.4) indicated maximum absorbance at 445 nm and the emission maximum at 575 nm (Fig. 5.7). Compared to the photophysical properties of 7-diethylaminocoumarin ($\lambda_{ex} = 410$ nm, $\lambda_{em} = 475$ nm), **NCC** showed bathochromic

Fig. 5.7 Absorbance (dashed) and emission (solid) spectra of 7-diethylamino coumarin (black, 20 μM, λ_{ex} = 405 nm) and **NCC** (red, 20 μM, λ_{ex} = 450 nm) in HEPES buffer (100 mM, pH 7.4)

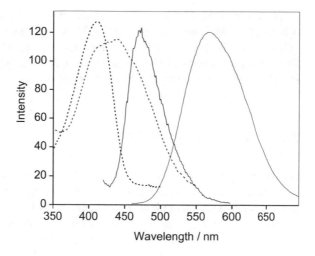

Fig. 5.8 Reduction of **NCC** (λ_{ex} = 445 nm). Untreated probe (black), **NCC** treated with 2000 molar equivalents of $Na_2S_2O_4$ (blue), $NaBH_3CN$ (red) and $NaBH_4$ (green). All spectra were acquired using 20 μM solution of probe in HEPES buffer (100 mM, pH 7.4)

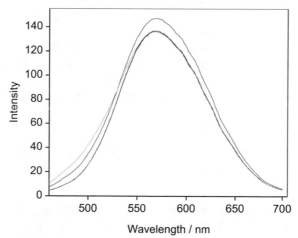

shifts of 35 nm in absorbance and 100 nm in fluorescence emission. This result confirmed the original hypothesis that the electron withdrawing property of the oxidised from of nicotinamide scaffold would result in a red-shift in the photophysical properties of the coumarin moiety.

Preliminary experiments were performed to measure changes in fluorescence properties of **NCC** upon treatment with reducing agents. However, it was observed that **NCC** did not show any change in its photophysical properties in the presence of large mole equivalents of mild reducing agents such as $Na_2S_2O_4$ and $NaBH_3CN$ (Fig. 5.8). It was initially postulated that extending the conjugation of the nicotinamide scaffold could have decreased its reduction potential, but even upon treatment with the strong reducing agent $NaBH_4$, the fluorescence properties of **NCC** did not change (Fig. 5.8).

A survey of the literature showed that all redox responsive structures based on nicotinamide, including the cellular NAD^+/NADH, possess a quarternary pyridinium cation [5, 6]. To investigate if the lack of this structural feature in **NCC** was responsible for its unresponsive behaviour towards changes in redox state, the structure of **NCC** was altered to incorporate a pyridinium cation.

5.2.3 Synthesis of NCR3

A quarternary nitrogen atom on the nicotinamide scaffold can be installed by alkylating the nitrogen atom in the heterocycle upon reaction with an alkyl halide prior to its condensation with the coumarin aldehyde. Considering the unsuccessful condensation of 5-methylnicotinamide (**51a**) with the coumarin aldehyde (**48**) in Sect. 5.2.1, this synthetic scheme was only performed with 6-methylnicotinamide (**51b**). N-Alkylation of **51b** with excess ethyl iodide in MeCN under reflux resulted in the formation of the corresponding pyridinium salt (**52**) (Scheme 5.3). **52** was then taken forward for condensation with **48** in DMF and p-toluenesulfonic acid to give nicotinamide-coumarin redox sensor-3 (**NCR3**) in 42% yield. Moreover, the higher yield obtained upon the condensation of **48** with **52** compared to **51b**, is a possible consequence of the enhanced activation of methyl group by the stronger electron withdrawing effect of the pyridinium cation in **48** compared to pyridine in **51b**.

5.2.4 Photophysical and Redox Responsive Properties of NCR3

The photophysical properties were studied using a 20 μM solution of **NCR3** in 100 mM HEPES buffer (pH 7.4). The maximum absorbance of **NCR3** was found to be 515 nm, with maximum emission at 645 nm (Fig. 5.9). This indicated significantly larger bathochromic shifts of 105 nm and 170 nm in absorbance and fluorescence respectively compared to 7-diethylaminocoumarin ($\lambda_{ex} = 410$ nm, $\lambda_{em} = 475$ nm). Owing to the presence of the highly electron withdrawing pyridinium cation, the redshift in the absorbance and emission of **NCR3** was much higher than that observed in case of **NCC**. Furthermore, because red light is known to have better penetration ability, **NCR3**, which has emission maximum lying in the red region of the visible spectrum (645 nm) could potentially be utilised in deep-tissue and animal imaging experiments.

The redox responsive properties of **NCR3** were interrogated by recording the spectral properties of the probe (20 μM, 100 mM HEPES, pH 7.4) in the presence of reducing and oxidising agents. Upon reduction with 0–50 molar equivalents of the

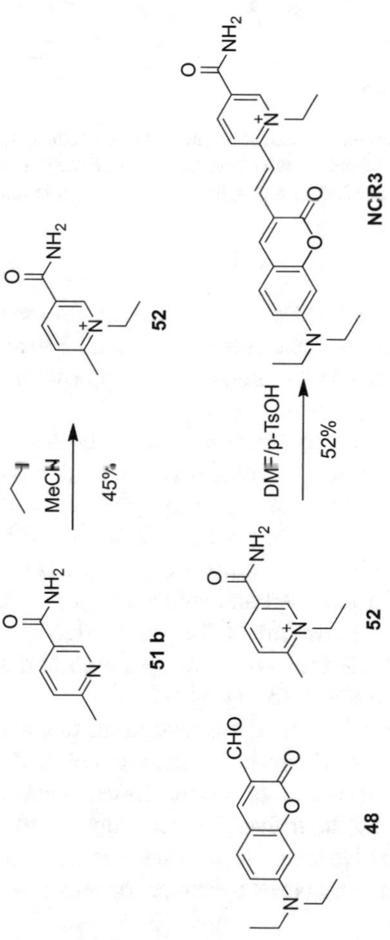

Scheme 5.3 Synthesis of **NCR3**

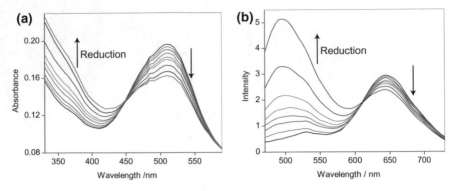

Fig. 5.9 Photophysical properties **a** absorbance and **b** fluorescence spectra ($\lambda_{ex} = 456$ nm) of **NCR3** (20 μM) in the oxidised form followed by reduction with incremental additions of NaBH$_3$CN (0–50 molar equivalents) as the reducing agent. Spectra were collected in HEPES buffer (100 mM, pH 7.4)

mild reductant NaBH$_3$CN, **NCR3** showed a decrease in the absorbance at 520 nm and a concomitant increase in the absorbance around 330 nm (Fig. 5.9a). In addition, the isosbestic point at 451 nm suggests the occurrence of only one chemical transformation, a reduction process.

Fluorescence measurements were recorded using the isosbestic point (451 nm) as the excitation wavelength. A similar trend was observed in the fluorescence spectra of **NCR3** upon reduction. The intensity of the peak at 645 nm decreased, accompanied by an increase in the peak at 500 nm (Fig. 5.9b), with an isoemissive point at 605 nm. The ratio of the intensity of the emission peaks can therefore be used to decipher the proportions of oxidised and reduced forms of **NCR3** and thus obtain information about its redox environment. The diminution of the peak at 645 nm can be attributed to the weaker electron withdrawing ability of the pyridine ring and the loss of conjugation in the reduced form (Fig. 5.5).

Unfortunately, addition of H$_2$O$_2$, at concentrations ranging from 20–1000 molar equivalents, did not restore the fluorescence spectrum of **NCR3**. This suggested that **NCR3** undergoes irreversible reduction and therefore cannot be chemically oxidised. One possible explanation for the irreversible reduction of **NCR3** could be chemical reduction of the vinylic bond joining the nicotinamide and coumarin scaffolds which could be further investigated by NMR spectroscopy and mass spectrometry [7].

5.3   Electrochemical Studies

Electrochemical measurements were performed in the laboratory of Dr. Conor F. Hogan at La Trobe University, in order to determine the reduction potential of **NCR3** and devise strategies to address its irreversible reduction. Cyclic voltammograms were recorded from a 2 mM solution of **NCR3** in MeCN containing tetrabutylam-

Fig. 5.10 Cyclic voltammogram of **NCR3** (2 mM) showing the reduction and oxidation events over two cycles. The voltammograms were recorded in MeCN containing 0.1 M TBAPF$_6$ as supporting electrolyte. The arrow indicates the starting point and initial direction of potential sweep

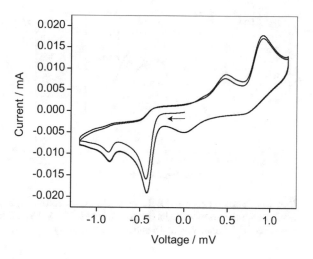

monium hexafluorophosphate (TBAPF$_6$, 0.1 M) as an electrolyte and ferrocene as an internal standard. As shown in Fig. 5.10, **NCR3** showed both reduction and oxidation events at potentials −448 and 932 mV respectively (versus Fc/Fc$^+$). The redox potential of the probe was calculated as −488 mV versus Fc/Fc$^+$. Such a high redox potential suggests that the reduced form of the probe is highly stable, and explains the irreversible reduction of **NCR3**. Nevertheless, spectro-electrochemical studies were performed to gain an understanding of the optical behaviour of **NCR3** in relation to its electrochemistry.

5.3.1 Spectro-Electrochemistry

Spectro-electrochemical studies were also performed at Dr. Conor F. Hogan's lab. A reduction potential of −0.7 V (versus Ag/Ag$^+$) was applied using a platinum gauze as the working electrode, platinum wire as the counter electrode and Ag/Ag$^+$ as the reference electrode. The absorbance and fluorescence spectra were recorded every 15 seconds following the application of the reduction potential. As shown in Fig. 5.11a, the peak at 500 nm decreased in intensity over subsequent scans accompanied with an increase in the peak at 380 nm, which is similar to the results obtained from chemical reduction of **NCR3** with NaBH$_3$CN. In addition, the isosbestic point obtained from this experiment (λ = 448 nm) is in close agreement with that obtained by chemical reduction (451 nm) confirming that the transformation process that occurs upon addition of NaBH$_3$CN is reduction.

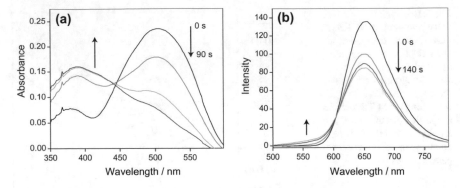

Fig. 5.11 **a** Absorbance and **b** fluorescence ($\lambda_{ex} = 458$ nm) spectra of **NCR3** (100 μM) over time after the application of a reduction potential of -0.7 V. The absorbance and fluorescence spectra were recorded at intervals of 15 and 20 s, respectively

The fluorescence spectra in Fig. 5.11b also show a decrease in the red fluorescence (645 nm), and a concomitant increase in the green fluorescence (550 nm). The increase in the peak at 550 nm was not as pronounced compared to that obtained from chemical reduction, which means that the emissive properties of the electrochemically reduced form are significantly compromised. Further investigations using bulk electrolysis might help understand such effects. Nevertheless, the values of the isoemissive point obtained from chemical and electrochemical reduction are in agreement.

The reversibility of the electrochemical and photophysical properties of **NCR3** were investigated to check whether applying an oxidising potential would show a complete reversal of the spectral pattern obtained in Fig. 5.11. This was tested by performing electrochemical reduction of the probe followed by an immediate application of a positive oxidation potential of 1.0 V (versus. Ag/Ag$^+$), with the photophysical properties recorded every 20 s (Fig. 5.12). As evidenced by the increase in absorbance at 500 nm and the associated decrease at 380 nm, applying an oxidation potential of 1.0 V resulted in the re-oxidation of **NCR3** (Fig. 5.12a). This is further confirmed by the fluorescence spectra, which show an increase in the peak intensity at 645 nm accompanied by a decrease in the peak at 550 nm (Fig. 5.12b). From these results it was understood that **NCR3** has immense potential for use as a reversible redox sensor. It was crucial to therefore be able to tune its structure such that the oxidation potential is much lower, and within the range of cellular oxidants (ROS).

At this stage, it was thought that suitable modifications on the nicotinamide scaffold, specifically the amide group at position 3, could help tune the redox potential of the probe and achieve a reversible redox response. Furthermore, the redox responsive properties of other moieties such as quinolinium group can be investigated. These

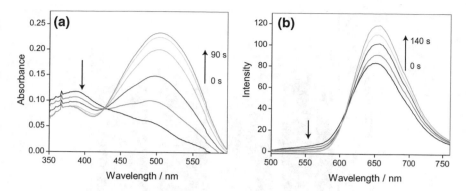

Fig. 5.12 **a** Absorbance and **b** fluorescence spectra of electrochemically reduced **NCR3** (100 μM) recorded after the application of an oxidation potential of 1.0 V. The absorbance and fluorescence spectra were recorded at intervals of 15 and 20 s, respectively

approaches were investigated by Madeleine Carr, an Honours student in the group, who developed quinolium-coumarin based redox probes which exhibited far-red fluorescence properties with excellent tissue penetration and ratiometric response to reduction. However, the challenge of achieving a reversible redox response remains.

5.4 Cell Permeability

Although **NCR3** was observed to possess irreversible redox behaviour, its ability to respond to reducing cellular environments was investigated. Confocal microscope images (λ_{ex} = 458 nm, λ_{em} = 470 − 700 nm) displayed no fluorescence from HeLa cells treated with **NCR3** (50 μM, 30 min), suggesting that **NCR3** cannot penetrate the cell membrane (Fig. 5.13). To investigate probe uptake in other cell types, treatments were repeated in DLD-1 cells and RAW 264.7 macrophages, but even with longer treatment times (1 and 2 h), the cells did not show any evidence of **NCR3** accumulation, with fluorescence similar to background levels.

In order to cross the phospho-lipid cell membrane by passive diffusion, a molecule must be sufficiently hydrophobic. It is therefore likely that the positive charge on the scaffold that was necessary for the redox response precluded the cellular penetration of **NCR3**. It was envisioned that the hydrophobicity of the molecule could be enhanced by having a longer alkyl chain on the pyridinium nitrogen. This hypothesis was tested by synthesising **NCR4** by the condensation of the same coumarin aldehyde (**48**), this time with 1-hexyl-6-methylnicotinamidium iodide (**53**), which was synthesised by alkylating **51b** with hexyliodide (Scheme 5.4).

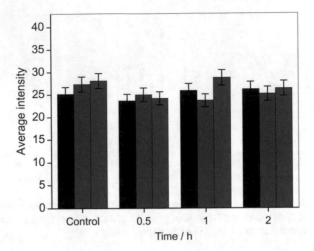

Fig. 5.13 Average intensity of images obtained by confocal microscopy of HeLa (black), DLD-1 (red) and RAW 264.7 (blue) cells when treated with vehicle control (DMSO) and **NCR3** (50 μM) for different durations. Bar graphs represent average intensities of three images. Error bars represent standard deviation $n = 3$

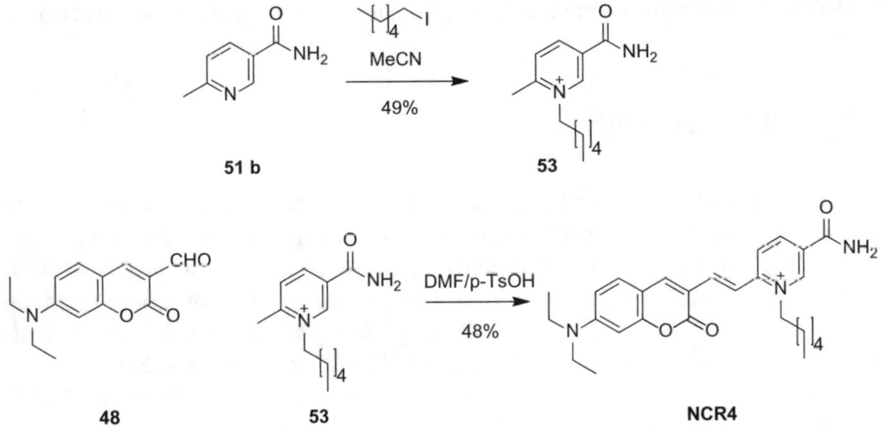

Scheme 5.4 Synthesis of **NCR4**

The photophysical and electrochemical behaviour of **NCR4** mimics that of **NCR3**. Following its purification and characterisation, the cell penetration ability of **NCR4** was interrogated by treating RAW 264.7 macrophages with **NCR4** (50 μM, 15 min). The confocal microscopy images ($\lambda_{ex} = 458$ nm, $\lambda_{em} = 470 - 700$ nm) showed the presence of bright red fluorescence within the cells (Fig. 5.14). Furthermore, the staining pattern was characteristic of mitochondrial localisation. However, at this stage, investigations to assess the ability of **NCR4** to image cellular oxidative capacity were limited by the irreversible reduction of **NCR4**.

Fig. 5.14 Confocal microscope images of RAW 264.7 cells treated with **NCR4** (50 μM, 15 min). Scale bar represents 20 μm

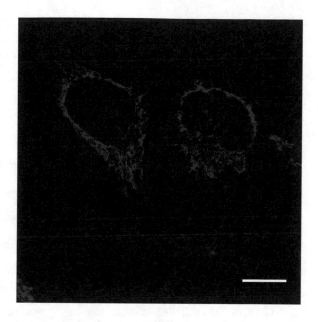

5.5 Conclusions

In designing ratiometric redox probes it is important to ensure that the reduction potential of the redox responsive component is not significantly altered upon conjugation to the fluorophore. This chapter investigated the utility of nicotinamide as a redox responsive moiety. The design strategies, synthesis and photophysical characterisation of nicotinamide based probes have been looked into. The results obtained from the photophysical studies of the developed molecules confirm ICT as a successful strategy to develop a red fluorescent ratiometric redox probe by conjugating nicotinamide and coumarin. However, in designing such probes the redox potential of the nicotinamide scaffold was pushed beyond the responsive window of common oxidising agents. Furthermore, when designing future redox probes based on nicotinamides, it is important to conserve the quarternary pyridinium cation within the probe structure to ensure redox responsive behaviour.

The spectro-electrochemical studies revealed immense potential of **NCR3** to perform as a reversible redox probe. However, future work would require suitable structural modifications to draw the reduction potential of the scaffold within an accessible range. In addition, isolation of reduced forms of **NCR3** and **NCR4** and their structural characterisation by NMR and mass spectroscopy would further aid is understanding the differences in chemical and electrochemical reduction. A simple switch of shorter alkyl chain with a longer one, helped address the cell permeability issues, highlighting the importance of lipophilicity in probe design toguarantee successful cell penetration. Moreover, the mitochondrial localisation of the lipophilic **NCR4** demonstrates great potential of the developed molecules to act as starting points for further investigations leading to the design of a mitochondrially targeted ratiometric redox probe.

References

1. P. Belenky, K.L. Bogan, C. Brenner, NAD+ metabolism in health and disease. Trends Biochem. Sci. **32**, 12–19 (2007)
2. F.L. Rodkey, J.A. Donovan, Oxidation-reduction potentials of the triphosphopyridine nucleotide system. J. Biol. Chem. **234**, 677–680 (1959)
3. R.D. Braun, An electrochemical study of flavin adenine dinucleotide. J. Electrochem. Soc. **124**, 1342–1347 (1977)
4. N.A. Kuznetsova, O.L. Kaliya, The photochemistry of coumarins. Russ. Chem. Rev. **61**, 683 (1992)
5. C.O. Schmakel, K.S.V. Santhanam, P.J. Elving, Nicotinamide adenine dinucleotide (NAD+) and related compounds. Electrochemical redox pattern and allied chemical behavior. J. Am. Chem. Soc. **97**, 5083–5092 (1975)
6. P. Leduc, D. Thévenot, Electrochemical properties of nicotinamide derivatives in aqueous solution: Part IV. Oxidation of N1-alkyl-1,4-dihydronicotinamides. J. Electroanal. Chem. Interfacial Electrochem. **47**, 543–546 (1973)
7. L. Tan, W. Lin, S. Zhu, L. Yuan, K. Zheng, A coumarin-quinolinium-based fluorescent probe for ratiometric sensing of sulfite in living cells. Organ. Biomol. Chem. **12**, 4637–4643 (2014)

Chapter 6
In Cellulo Studies

Understanding the morphological and physiological processes that occur within a biological system is critical to addressing many questions pertaining to health and disease. Cell culture refers to the isolation of cells from animals and subsequent maintenance in a favourable artificial environment. It remains one of the simplest models for investigating biological systems, free of the complications that arise from studying tissue or whole animals [1]. With cell culture, it is possible to study a limited numbers of cells in a well-defined environment. Furthermore, the physico-chemical (oxygen concentration, temperature and pH) and physiological (hormone, serum and nutrient concentrations) aspects of the environment in which the cells propagate can be manipulated as desired, to mimic specific conditions, whether the onset of a pathological condition [2], exposure to pathogens [3] or to understand the effect of environmental variations on basic cellular processes [4–6]. In addition, the genetic make-up of the cells can be precisely controlled to obtain deeper understanding of effect of gene mutations and their role in causing various hereditary diseases [7].

Current life science research employs cultured cells in a wide range of investigations, such as evaluating the therapeutic and/or cytotoxic effects of drugs [3], interactions between cell and pathogens [8], and understanding the processes such as mutagenesis, carcinogenesis and aging [9]. This has been achieved by a variety of methods, such as biochemical assays [10], blotting techniques [11] and optical imaging [12]. This chapter details the experiments performed using the fluorescent redox probes developed during this PhD (Fig. 6.1), to image variations in the oxidative capacity of a range of cultured cells.

6.1 Embryonic Stem Cells

Embryonic stem cells (ESCs) procured from the inner cell mass of the embryonic blastocyst are pluripotent, meaning that they have the ability to differentiate into many

© Springer International Publishing AG 2018
A. Kaur, *Fluorescent Tools for Imaging Oxidative Stress in Biology*,
Springer Theses, https://doi.org/10.1007/978-3-319-73405-7_6

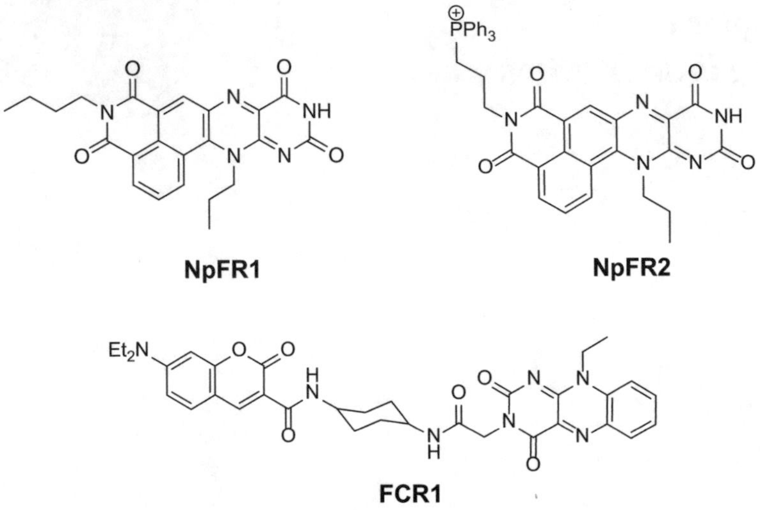

Fig. 6.1 Chemical structures of the redox probes utilised in *in cellulo experiments*

specialised cell types [13]. The pluripotency of embryonic cells has been reported to be dependent on a variety of intrinsic regulators and the exogenous environment. Evidence suggests that extracellular factors with specific cell surface receptors initiate diverse intracellular signal transduction pathways [14]. Within a cell, secondary messengers play a crucial role in the amplification of the signalling pathways responsible for diverse cellular responses, including proliferation, differentiation and apoptosis [15], and the role of ROS/RNS as secondary messengers has recently been demonstrated [14].

Several recent reports have successfully established the role of ROS in facilitating the differentiation of ESCs into specific cell types [16–19], and minute variations in the ROS-levels in ESCs have been shown to have drastic effects on stem cell fate [20–22]. Furthermore, recent studies highlight the fact that ROS can induce the differentiation of ESCs into the mesoderm (the middle layer of cells in an embryo) and endoderm (the inner layer of cells in an embryo). However, the expression of ectoderm (the outer layer in an embryo)-related genes remained unaltered [23]. In addition, particularly in embryonic stem cells, it has been demonstrated that fluxes in the cellular oxidative capacity play a key role in the mediating the mitochondria-nucleus crosstalk which could be responsible for establishing a coordination between metabolic and differentiation events of the cell [24, 25]. This section details the investigations that have been performed using **NpFR2** to explore the variations in mitochondrial ROS-levels in embryonic stem cells and their effect on the stem cell differentiation process.

6.1.1 Imaging Redox State in Embryonic Stem Cells

Prior to interrogating the role of cellular redox state in mediating stem cell differentiation, preliminary investigations were performed to investigate the ability of **NpFR2** to respond to changes in the mitochondrial oxidative capacity of ESCs brought about by exogenous redox agents. These studies were performed in collaboration with Dr. Stuart Fraser and Mr. Kurt Brigden in the School of Medical Sciences, University of Sydney.

To mimic reducing and oxidising cellular environments, murine ESCs were treated with reducing (NAC, $50 \mu M$, 30 min) or oxidising agents (H_2O_2, $50 \mu M$, 30 min) followed by treatment with **NpFR2** ($20 \mu M$, 15 min). The cells were then washed and prepared for assay on a flow cytometer. Basal oxidative capacity was measured using cells treated only with **NpFR2** ($20 \mu M$, 15 min), and unstained control cells treated with vehicle control (DMSO) were also analysed. Analysis of flow cytometry results illustrated that ESCs treated with the probe had significantly greater fluorescence intensity compared to the unstained control (Fig. 6.2). The cells treated with H_2O_2 exhibited 4-fold higher **NpFR2**-fluorescence intensity compared to cells treated with probe alone, corresponding to higher mitochondrial oxidative stress in these cells, as expected.

Furthermore, the fluorescence intensity of cells treated with the reducing agent NAC was significantly reduced compared to the oxidised cells and those treated with probe alone (Fig. 6.2). In fact, the fluorescence profile of reduced cells was observed to be closer to that of unstained cells. The results illustrate that **NpFR2** provides a clear readout of the mitochondrial redox state of ESCs.

6.1.2 Role of Copper in Cellular Redox Homoeostasis

Copper is an essential micro-nutrient utilised by all living organisms for growth and development. Possessing a high redox potential, copper operates as a cofactor for both enzymatic and non-enzymatic proteins involved in a variety of biological redox reactions [26, 27]. Owing to the preponderance of electron acceptors such as molecular oxygen and superoxide radical anion ($O_2^{\bullet -}$) intracellular copper is oxidised from cuprous to cupric species [28]. Facilitated by its redox-active biochemistry, copper participates in non-enzymatic Haber-Weiss (Eq. 6.1) and Fenton (Eq. 6.2) reactions that generate ROS [29, 30].

$$(O_2^{\bullet -}) + Cu^{2+} \rightarrow O_2 + Cu^+ \tag{6.1}$$

Fig. 6.2 Flow cytometric studies of murine embryonic stem cells treated with **NpFR2** (20 μM, 15 min, λ_{ex} = 488 nm) showing the fluorescence intensity of cells treated with vehicle control, probe alone, *N*-acetyl cysteine, and H_2O_2

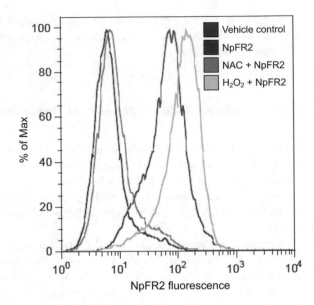

$$H_2O_2 + Cu^+ \rightarrow OH^- + {}^\bullet OH + Cu^{2+} \tag{6.2}$$

To attenuate the pathological consequences of excess labile copper, a set of sophisticated and highly-conserved copper transporters and metallo-chaperones have evolved that play a crucial role in transporting copper to explicit intracellular targets and in maintaining copper homoeostasis [31, 32]. For example, CTR1 is a ubiquitously expressed plasma membrane protein composed of 3 transmembrane domains that form a homotrimeric pore with high affinity for the Cu^+ ion [32]. CTR1 shuttles Cu^+ across the membrane and has been demonstrated to play a key role in maintaining copper homoeostasis particularly during embryogenesis and brain development [33].

CTR1 knock-out mice have been created, and their phenotype is instructive of the protein's role in brain development and copper homoeostasis [33, 34]. Homozygous CTR1 knock-out embryos (CTR1$^{-/-}$), which contain no functional copy of CTR1 gene (SLC31A1), do not survive beyond 8.5 *days post coitum* (dpc, referring to days after copulation) due to significant developmental abnormalities [33]. In contrast, CTR1 heterozygous embryos (CTR1$^{+/-}$), which contain one functional copy of the CTR1 gene are viable, and commence organogenesis 8.5 dpc, similar to their wild-type (CTR1$^{+/+}$) littermates [33]. It is understood that several cupro-enzymes facilitate key protein-transformations during gastrulation, so copper deficiency could be a crucial factor compromising the development of embryonic germ layers.

6.1.3 Mitochondrial Redox State in CTR1 Knock-Outs of Embryonic Stem Cells

Considering the role of copper in regulating mitochondrial-ROS levels *via* the Haber-Weiss and Fenton-like chemistry, and the role of ROS as signalling molecules during mesendodermal differentiation of ESCs, it was hypothesised that one possible mechanism by which CTR1$^{-/-}$ fail to survive embryogenesis could be putatively low ROS levels in these embryos. In order to confirm this hypothesis, investigations were performed to measure the mitochondrial ROS levels in three murine embryonic stem cell lines expressing varying degrees of the copper transporter (CTR1$^{+/+}$ cells, CTR1$^{+/-}$ and CTR1$^{-/-}$). These cell lines were maintained by Mr. Kurt Brigden, a member of the Fraser lab at the School of Medical Sciences, University of Sydney.

Having demonstrated the ability of **NpFR2** to evaluate mitochondrial redox state in ESCs, the probe was employed for this investigation. Three murine ESC lines—CTR1$^{+/+}$ cells, CTR1$^{+/-}$ and CTR1$^{-/-}$ were treated with **NpFR2** (20 μM, 15 min) and the fluorescence emission was analysed by flow cytometry. The population of wild type CTR1 ESCs (CTR1$^{+/+}$) exhibited the highest fluorescence intensity compared to the heterozygotes (CTR1$^{+/-}$), whilst the least fluorescence intensity was observed in case of the CTR1 homozygous knock-outs (CTR1$^{-/-}$) (Fig. 6.3).

These results indicate that there are considerably reduced levels of mitochondrial ROS in the CTR1$^{-/-}$ ESCs compared to the wild type CTR1$^{+/+}$ cells, thus confirming our hypothesis that reduced cellular copper levels translate into lower mitochondrial oxidative capacity in the cells.

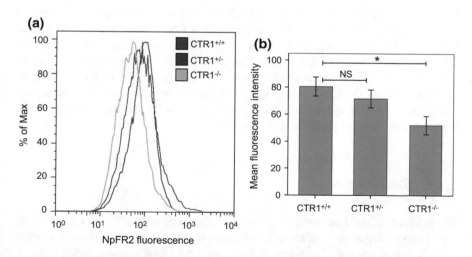

Fig. 6.3 Flow cytometric studies of murine embryonic stem cells expressing varying levels of CTR1 treated with **NpFR2** (20 μM, 15 min, λ_{ex} = 488 nm). **a** Representative histograms of cell populations, **b** mean fluorescence intensities of cell populations. Error bars represent standard deviation (n = 4), $p < 0.05$ is considered significant

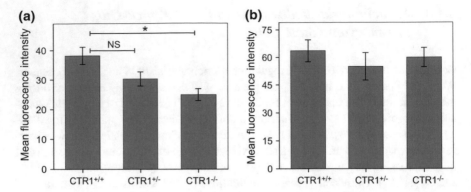

Fig. 6.4 Flow cytometric studies of Murine embryonic stem cells expressing varying levels of CTR1 when subjected to (**a**) mesodermal and **b** ectodermal differentiation conditions treated with **NpFR2** ($20\,\mu M$, $15\,min$, $\lambda_{ex} = 488$ nm). Bars represent mean fluorescence intensities of cell populations. Error bars represent standard deviation ($n = 3$), $p < 0.05$ is considered significant

Next, investigations focussed on assessing how mitochondrial ROS levels changes during embryogenesis of CTR1 knock-out ESCs were performed. In order to achieve this, the three ESC lineages were exposed to conditions facilitating mesodermal and ectodermal differentiation in an attempt to mimic in utero embryogenesis. These cells were allowed to differentiate in vitro over a period of 5 days, after which they were dosed with **NpFR2** and analysed on a flow cytometer. Interestingly, no significant variations in fluorescence intensity were observed during the ectodermal differentiation of CTR1$^{+/+}$ cells, CTR1$^{+/-}$ and CTR1$^{-/-}$ ESCs (Fig. 6.4) indicating unchanged mitochondrial ROS levels. However, under conditions of mesodermal differentiation, CTR1 $^{-/-}$ exhibited a much lower fluorescence intensity compared to the CTR1$^{+/+}$ cells, whilst no significant differences were observed between the CTR1 $^{+/-}$ and CTR1 $^{+/+}$ ESCs. This illustrates that the mitochondrial ROS levels are not significantly lowered when at least one functional copy of the CTR1 gene is present within the ESC. However, in a homozygous knock-out, the mitochondrial ROS levels are compromised and could therefore be playing a crucial role in hampering the normal differentiation process.

6.1.4   Conclusions

The results obtained from these experiments indicate that complete knock-out of the CTR1 expressing gene is linked to lower levels of mitochondrial oxidative capacity. A similar trend was observed when the ESCs were subjected to mesodermal differentiation conditions, as inferred from the fluorescence intensity of **NpFR2**. However, **NpFR2** reported no significant variation in mitochondrial ROS levels upon exposure to ectodermal differentiation conditions, warranting further investigation to better understand this observation.

6.2 Pancreatic Cancer Cells

Mia PaCa-2 (MP) pancreatic cancer cells have been extensively used to investigate pancreatic cancer cell biology [35–38]. A variety of cancer cells including pancreatic, thyroid, ovary, breast and lung cancer have been found to express high levels of progesterone receptor membrane component 1 (PGRMC1) [39–41]. PGRMC-1 has been reported to contribute towards protection of the cancer cells from damage and promote their survival and proliferation. [42, 43]. The prime biochemical function of this protein, which shares key structural motifs with cytochrome b_{55}, is heme-binding [44, 45].

To investigate the potential roles of the PGRMC1-dependent signalling system in pancreatic cancer cells, Dr. Michael Cahill's research group at Charles Sturt University generated stable mutants with varying degrees of PGRMC1-phosphorylation by stably transfecting MP cells with a plasmid expressing PGRMC1 variants. The mutations involve replacing commonly phosphorylated amino acids such as serine and tyrosine with ones that do not undergo phosphorylation. The developed mutants include the wild-type (WT) sequence, a S57A/S181A double mutant (DM), in which serine amino acid at positions 57 and 181 are changed to alanine [46], and the S57A/Y180F/S181A triple mutant (TM), containing an additional mutation with tyrosine 180 replaced with phenylalanine. Preliminary data from Dr. Cahill's investigations using proteomic studies and metabolic assays suggested that varying PGRMC1 phosphorylation status could affect mitochondrial function, glucose uptake and lactate production in these cells. The aim of this section of the work was to investigate any effect PGRMC1-phosphorylation might have on the cellular and mitochondrial redox homoeostasis. MP cells and the three mutants were provided by Dr. Cahill.

6.2.1 Variations in Oxidative Capacity of PGRMC1 Mutants

The cytoplasmic oxidative capacities of MP, WT, DM and TM cells were interrogated and **NpFR1** was used for this purpose. These studies were performed by Mr. Partho Adhikary from the Cahill group at the Charles Sturt University [47]. The cells were treated with 20 μM of **NpFR1** and assayed for fluorescence by flow cytometry. **NpFR1** exhibited two cell populations: one with lower (peak A) and one with a higher (peak B) **NpFR1**-fluorescence intensity illustrating that these populations correspond to lower and higher cytoplasmic oxidative capacities respectively (Fig. 6.5).

Cellular redox state has been reported to differ between various cell cycle stages. The G2/S has a more reducing cellular environment and the S and G1 phases have intermediate and more oxidising environments respectively [48]. It is therefore suggested that peak A with lower oxidative capacity could potentially represent cells in G2/M phase, whilst the peak B could represent those in G1 phase.

Fig. 6.5 Mia PaCa-2 cells
treated with **NpFR1** showing
the presence of two different
population (peaks A and B)
of cells based on the intensity
of **NpFR1** fluorescence flow
cytometry results

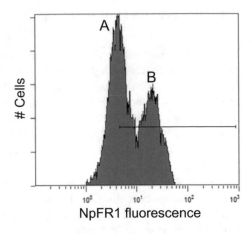

While no significant fluorescence was observed in untreated control cells, the results suggest that a greater proportion of MP cells constitute peak A (**NpFR1**-G2/M peak) compared to that of DM and WT cells. TM cells also exhibited 4-times fewer cells in peak A (**NpFR1**-G2/M peak) than other cell types (Fig. 6.6). These patterns were conserved across three different clones of each cell type. Furthermore, peak B of TM cells exhibited lower cytoplasmic oxidative capacity (**NpFR1**-G1 peak) compared to MP and DM cells (indicated by the left shifted peak B (**NpFR1**-G1 peak) against the average high fluorescence intensity (vertical dotted line, Fig. 6.6).

In order to assess the mitochondrial oxidative capacity, cells were treated with **NpFR2** and assayed for fluorescence by flow cytometry. The results indicate more than 90% of WT and TM cells lie within the high fluorescence gated, while approximately 75% of MP and DM cells lie in this gate. This suggests that the inner mitochondrial matrix of WT and TM cells was more oxidising compared to that of MP and DM cells (Fig. 6.7). An interesting observation was that the WT and TM cells which exhibited more oxidising mitochondrial matrix demonstrated a less oxidising cytoplasmic environment. To obtain a clearer picture of the mitochondrial function and stress responses of these cells, Seahorse XF Cell Mito Stress tests were performed at the Molecular Biology Facility, Bosch Institute, University of Sydney.

6.2.2   Assessing Mitochondrial Activity

Cellular metabolism involves the uptake of nutrients, such as oxygen, glucose and glutamine for energy production through a series of enzymatically regulated redox reactions. The Seahorse XF Cell Mito Stress Test measures the rate of cellular oxygen consumption, an indicator of cellular respiration (oxygen consumption rate, OCR measured as picomoles/minute), and proton excretion (extracellular acidification rate, ECAR measured as millipH/min) that represents the rate of glycolysis.

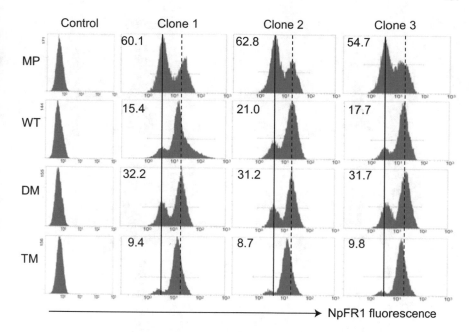

Fig. 6.6 NpFR1 flow cytometry results. The numbers at the upper left of each panel represent the percentage of cells in peak A (left peak). The vertical dotted lines represent the average fluorescence intensities for peak A (solid) and B (dotted) for MP and/or DM cells

In the Seahorse assay, the cellular metabolism is perturbed by the addition of three different metabolic modulators (Fig. 6.8). The first one is oligomycin, an inhibitor of ATP synthesis which works by blocking the proton channel of the Fo portion ATP synthase (Complex V). It can therefore be applied to differentiate between the percentage of oxygen consumption dedicated to ATP production and that essential to overcome the natural proton leak across the inner mitochondrial membrane (Fig. 6.8).

This is followed by injection of FCCP (carbonyl cyanide-*p*-trifluoromethoxy phenylhydrazone), an ionophore that disrupts the mitochondrial membrane potential by leaking protons across the membrane instead of the ATP synthase (Complex V) proton channel. The depolarised mitochondrial membrane potential results in accelerated consumption of energy and oxygen, without ATP production. The difference between the basal and maximal OCR would give the spare respiratory capacity, a major factor in determining cell vitality and survival (Fig. 6.8). The third injection is a cocktail of rotenone and antimycin A, inhibitors of Complex I and III respectively, which terminates mitochondrial respiration. Consequently, respiration resulting from both mitochondrial and non-mitochondrial fractions can be calculated. Therefore, key parameters of mitochondrial activity—basal respiration, ATP-linked respiration, maximal respiration, spare respiratory capacity, and non-mitochondrial respiration can be determined.

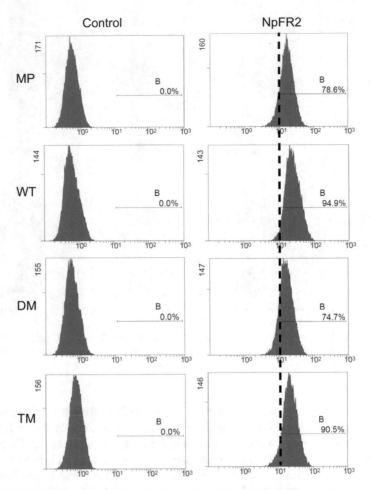

Fig. 6.7 **NpFR2** flow cytometry results. The numbers in each panel represent the percentage of cells to the right of the reference

6.2.2.1 Optimisation Assays

To effectively assess the bioenergetic profile of a cell type, it is essential to characterise its metabolic activity under basal and stress conditions. This is achieved by optimising the cell density and FCCP (ATP uncoupler) concentration for a given cell type.

According to the manufacturer's guidelines, OCR values between 50–400 pmol/min and ECAR values between 20–120 mpH/min are recommended for optimal cell density. To determine the optimal cell density, MP cells were plated at densities of 2×10^4, 4×10^4, 8×10^4 and 16×10^4 cells per well on a 24-well Seahorse cell culture plate and allowed to adhere overnight. Prior to the assay, the cells were

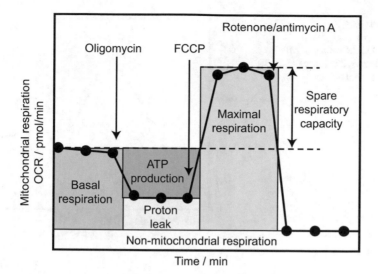

Fig. 6.8 A typical Seahorse mitochondrial stress test profile indicating the key parameters of mitochondrial activity Image source Seahorse Bioscience, CA, USA

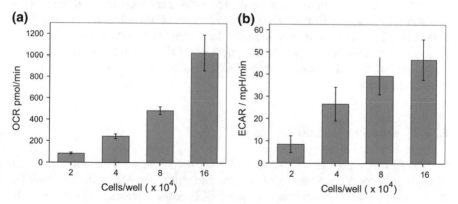

Fig. 6.9 Cell density optimisation assay. The basal **a** OCR and **b** ECAR values of MP cells plated at different cell densities

washed and incubated in serum-free XF assay medium (low-buffered, carbonate-free medium supplemented with 10 mM glucose, 2 mM sodium pyruvate and 1 mM L-glutamine) for 1 h. The basal OCR and ECAR were measured using a Seahorse XF 24 analyser. The OCR values of 2×10^4 (84 pmol/min) and 4×10^4 (244 pmol/min) cells/well lie within the recommended values (Fig. 6.9a). However, the basal ECAR for 2×10^4 cells/well was measured to be 8.5 mpH/min, which is below the recommended values (Fig. 6.9b). Therefore, the cell density of 4×10^4 cells/well with an ECAR of 26 mpH/min (Fig. 6.9b), was determined to be the optimal cell density for the assay.

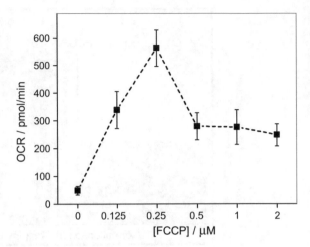

Fig. 6.10 FCCP optimisation assay. The maximal OCR values of MP cells treated with incremental FCCP concentrations

The optimum concentration of FCCP was determined using MP cells plated at the optimal density of 4×10^4 cells/well, which were washed and incubated in XF assay medium for 1 h before the assay. The cells were then injected with FCCP concentrations ranging from 0–2 μM. According to the manufacturer's protocol, the lowest FCCP concentration which yields the highest OCR (indicative of maximal respiration) is considered optimal. The highest OCR value was obtained for cells treated with 0.25 μM concentration of FCCP (Fig. 6.10) and was therefore used in subsequent assays.

6.2.2.2   Mitochondrial Stress Assays

In order to investigate the effects of PGRMC1 mutations on mitochondrial bioenergetics, mitochondrial stress assays were performed for MP, WT, DM and TM cells, using the optimal cell density and FCCP concentrations determined in the previous section. 4×10^4 cells/well were plated and allowed to adhere overnight, following which the cells were washed and incubated in XF assay medium for 1h. During the assay, the basal OCR was measured over three mix/wait/measure cycles—3 min/2 min/3 min. This was followed by sequential treatment with oligomycin (1μM), FCCP (0.25 μM) and a mixture of rotenone/anitmycin A (0.5 μM each). The concentrations of oligomycin and rotenone/antimycin A were used according to the manufacturer's guidelines. Following each treatment the OCR was measured over four mix/wait/measure cycles.

The mitochondrial stress profiles (Fig. 6.11a) demonstrate similar levels of basal respiration in all the cell types. However, significant differences were observed in the levels of maximal respiration (and therefore the spare respiratory capacity) between the cell types.

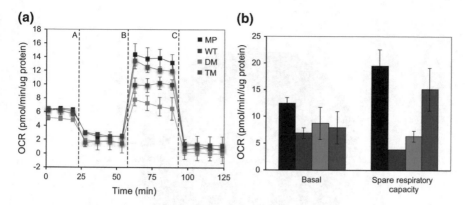

Fig. 6.11 **a** Mitochondrial stress profiles of MP cells and the PGRMC1 mutants. Dotted lines indicate the injection times of (A) oligomycin. (B) FCCP and (C) rotenone/antimycin A. **b** Bar graphs showing the OCR under basal conditions and the spare respiratory capacity of the cells. The OCR values are normalised to the protein content in each well. Error bars represent standard deviation $n = 3$

Both MP and TM cells showed greater spare respiratory capacity compared to WT and DM cells (Fig. 6.11b). Spare respiratory capacity is an important indicator of the bioenergetics of a cell and reflects the cells' ability to respond to an increase in energy demand by upregulating the mitochondrial electron transport chain [49, 50]. Lower spare respiratory capacity could indicate a mitochondrial dysfunction which might not be manifest under basal conditions, but becomes evident in conditions of high ATP demand [49, 50].

These results indicate that PGRMC1 mutation results in some level of mitochondrial dysfunction in the DM cells that is alleviated in the TM cells, which is in agreement with preliminary proteomics data obtained by the Cahill group. However, further experiments interrogating the variations in mitochondrial number and morphology using high resolution imaging would help understand the bioenergetic profiles of these cells.

6.2.3 Conclusions

The experiments with MP cells and mutants expressing varying degrees of PGRMC1 mutations suggest that the mutations significantly alter both cytoplasmic and mitochondrial oxidative capacity. In particular, the mitochondrial environment of WT and TM cells was observed to be highly oxidising compared to MP and DM cells. In contrast, the cytoplasmic redox state of WT and TM cells was less oxidising. Mitochondrial stress assays indicate that the mitochondria in the DM and WT cells exhibit lower spare respiratory capacity which could possibly be linked to dysfunctional mitochondria. However, further investigations are imperative to understanding the observed variations in the mitochondrial respiration profiles.

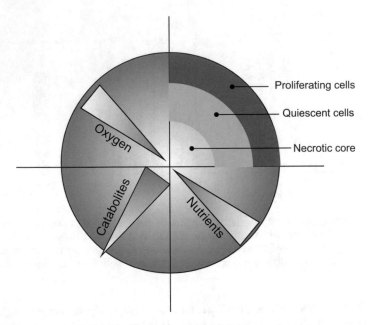

Fig. 6.12 The micorenvironment in a spheroid contains gradients of nutrient and oxygen availability as well as accumulation of toxic catabolites, and zones with distinct cell physiology

6.3   Three Dimensional Tumour Spheroids

Tumour spheroids are three dimensional spherical aggregates of cells formed by self-assembly [7]. Unlike classical 2-dimensional cell monolayers, spheroids mimic the 3D cellular context and tissue complexity with characteristic pathophysiology of in vivo tumours, making them an ideal model for the study of mechanisms involved in the onset and progression of cancer and the screening of potential therapeutic agents [51, 52]. The architecture of spheroids is strikingly similar to the in vivo tumour microenvironment exhibiting gradients of nutrient and oxygen availability as well as accumulation of toxic catabolites (Fig. 6.12). Spheroids with a radius of 200–500 µm can be classified into three zones—the outermost zone constitutes mainly of proliferating cells, followed by a zone of quiescent cells on the inside which thrive in conditions of limited nutrient and oxygen transport limitations, and the core of a spheroid usually harbours necrotic cells which reside in severely hypoxic conditions [53, 54].

As spheroids increase in size, concentration gradients of nutrients, oxygen, catabolites and other factors develop. Tumour microenvironments such as hypoxia have been demonstrated to contribute towards determining the growth kinetics and drug susceptibility of tumours as well as inducing necrosis [55, 56]. Common methods of determining oxygen gradients within tumour spheroids involve the use of microelectrodes [57, 58], analysing the expression levels of hypoxia-responsive protein

[59] and metabolism of radiolabelled substances [60, 61] most of which are invasive. Recently, optical imaging agents based on transition metal complexes have been developed and possess properties responsive to hypoxic environments [60, 62].

Considering the fact that the primary source of ROS/RNS is oxygen, *via* leakage of electrons from the mitochondrial electron transport chain, it was envisioned that a gradient in oxygen availability present between the core and periphery of a spheroid could give rise to gradient in oxidative capacity of cells in a tumour spheroid. To investigate this hypothesis, the ratiometric redox probe **FCR1** was employed for imaging 3D spheroids.

Spheroids were cultured from DLD-1, human colorectal adenocarcinoma cells. Approximately 25000 cells were plated on a 96-well plate containing a pad of $40\,\mu L$ agarose (750 mg in 100 mL PBS) in each well and incubated for a period of 4 days allowing the cells to assemble into a tight spheroid approximately 200–$300\,\mu m$ in radius. The spheroids were then carefully picked up using a pipette and placed onto a glass-bottom dish containing advanced DMEM supplemented with $20\,\mu M$ **FCR1** and incubated for a period of 16 h following which the spheroids were washed and prepared for imaging.

The images were obtained using multi-photon confocal microscope with excitation at 820 nm. The images recorded are consistent with accumulation of **FCR1** throughout the whole spheroid (Fig. 6.13). The images were obtained in the blue and green emission channels and a ratio image was developed using the RatioPlus plugin in FIJI (National Institute of Health). Although it was expected that spheroids would present a gradient of oxidative capacity, decreasing from periphery towards the core, the obtained ratio image demonstrated an interesting profile of cellular oxidative capacity. As shown in Fig. 6.13, the spheroid exhibited an intermediate ring of cells having high oxidative capacity, whilst the inner necrotic core and outermost layer of cells demonstrated comparatively lower levels of oxidative capacity. This result suggests that quiescent cells in the zone of nutrient and oxygen scarcity are oxidatively stressed compared to the proliferating outer core and hypoxic inner core.

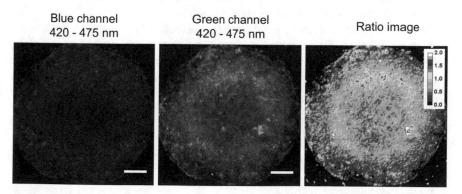

Fig. 6.13   Confocal microscopy images of DLD-1 spheroids treated with **FCR1** ($20\,\mu M$, 16 h, λ_{ex} = 820 nm) in blue and green channels. Scale bar represents $100\,\mu m$

Fig. 6.14 Fluorescence
lifetime imaging microscopy
of of DLD-1 spheroids
treated with **FCR1** (20 μM,
16 h, λ_{ex} = 820 nm)
illustrating different zones of
donor lifetimes at different
depths of the spheroid (zone
A and C, τ_m = 2.0 ns, and
zone B τ_m = 1.7 ns)

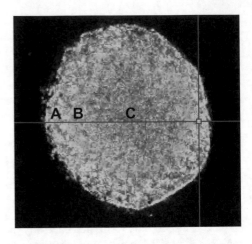

Similar results were obtained from lifetime imaging studies performed with
FCR1-treated DLD-1 spheroids described in Sect. 3.4.7, the donor lifetimes were
imaged in TCSPC mode. As shown in Fig. 6.14, an intermediate ring of cells having shorter donor lifetimes (1.7 ns) was observed, suggesting a greater FRET efficiency and consequently higher oxidising environment. In agreement with the results
obtained from confocal microscopy, the peripheral and inner-core cells demonstrated
longer donor lifetimes (2.0 ns), which relates to the poorer FRET efficiency and a
comparatively lower oxidative capacity.

6.3.1 Conclusions

Three dimensional spheroids present a valuable model to investigate tumour microenvironments in laboratory conditions. The results obtained confirm that the potential
of **FCR1** to report on cellular redox state is not just limited to cell monolayers, but
can be expanded to three dimensional tumour spheroids. These experiments demonstrate that the intermediate zone comprises of oxidatively-stressed quiescent cells,
whilst the cells in the core and close to the surface of the spheroid have comparatively lower levels of oxidative stress. Further investigations could be performed in
spheroids formed from cells expressing hypoxia responsive element conjugated to a
fluorescent protein, to better understand the hypoxic gradient within a 3-D tumour
spheroid.

References

1. J. Paul, *Cell and Tissue Culture* (Churchill Livingstone, 1975)
2. H.A. Laken, M.W. Leonard, Understanding and modulating apoptosis in industrial cell culture. Curr. Opin. Biotechnol. **12**, 175–179 (2001)
3. S. Ozturk, W.S. Hu, *Cell Culture Technology for Pharmaceutical and Cell-Based Therapies (Biotechnology and Bioprocessing)* (CRC Press, New York, 2005)
4. S.P. Langdon, *Cancer Cell Culture: Methods and Protocols*. Methods in Molecular Medicine (Humana Press, 2004)
5. M.P. Lutolf, P.M. Gilbert, H.M. Blau, Designing materials to direct stem-cell fate. Nature **462**, 433–441 (2009)
6. J.B. Griffiths, Serum and growth factors in cell culture media-an introductory review. Dev. Biol. Stand. **66**, 155–160 (1987)
7. T. Lai, Y. Yang, K.S. Ng, *Advances in Mammalian Cell Line Development Technologies for Recombinant Protein Production* (2013)
8. K.E. Hellstrom, I. Hellstrom, Immunological enhancement as studied by cell culture techniques. Annu. Rev. Microbiol. **24**, 373–398 (1970)
9. T. Reya, S.J. Morrison, M.F. Clarke, I.L. Weissman, Stem cells, cancer, and cancer stem cells. Nature **414**, 105–111 (2001)
10. G. Wu, *Assay Development: Fundamentals and Practices* (Wiley, 2010)
11. K. Wilson, J. Walker, *Principles and Techniques of Biochemistry and Molecular Biology*. Principles and Techniques of Biochemistry and Molecular Biology (Cambridge University Press, 2010)
12. M. Aswendt, J. Adamczak, A. Tennstaedt, A review of novel optical imaging strategies of the stroke pathology and stem cell therapy in stroke (2014)
13. M.K. Carpenter, E.S. Rosler, G.J. Fisk, R. Brandenberger, X. Ares, T. Miura, M. Lucero, M.S. Rao, Properties of four human embryonic stem cell lines maintained in a feeder-free culture system. Dev. Dyn.: An official publication of the American Association of Anatomists **229**, 243–258 (2004)
14. S.G. Rhee, Redox signaling: hydrogen peroxide as intracellular messenger. Exp. Mol. Med. **31**, 53–59 (1999)
15. R.S. Sohal, R.G. Allen, C. Nations, Oxygen free radicals play a role in cellular differentiation: an hypothesis. J. Radic. Biol. Med. **2**, 175–181 (1986)
16. B. Ateghang, M. Wartenberg, M. Gassmann, H. Sauer, Regulation of cardiotrophin-1 expression in mouse embryonic stem cells by HIF-1alpha and intracellular reactive oxygen species. J. Cell Sci. **119**, 1043–1052 (2006)
17. L. Ding, X.-G. Liang, Y. Hu, D.-Y. Zhu, Y.-J. Lou, Involvement of p38MAPK and reactive oxygen species in icariin-induced cardiomyocyte differentiation of murine embryonic stem cells in vitro. Stem Cells Dev. **17**, 751–760 (2008)
18. S. Lange, J. Heger, G. Euler, M. Wartenberg, H.M. Piper, H. Sauer, Platelet-derived growth factor BB stimulates vasculogenesis of embryonic stem cell-derived endothelial cells by calcium-mediated generation of reactive oxygen species. Cardiovasc. Res. **81**, 159–168 (2009)
19. H. Sauer, M. Wartenberg, Reactive oxygen species as signaling molecules in cardiovascular differentiation of embryonic stem cells and tumor-induced angiogenesis. Antioxid. Redox Signal. **7**, 1423–1434 (2005)
20. K. Ito, A. Hirao, F. Arai, S. Matsuoka, K. Takubo, I. Hamaguchi, K. Nomiyama, K. Hosokawa, K. Sakurada, N. Nakagata, Y. Ikeda, T.W. Mak, T. Suda, Regulation of oxidative stress by ATM is required for self-renewal of haematopoietic stem cells. Nature **431**, 997–1002 (2004)
21. K. Ito, A. Hirao, F. Arai, K. Takubo, S. Matsuoka, K. Miyamoto, M. Ohmura, K. Naka, K. Hosokawa, Y. Ikeda, T. Suda, Reactive oxygen species act through p38 MAPK to limit the lifespan of hematopoietic stem cells. Nat. Med. **12**, 446–451 (2006)
22. Y.-Y. Jang, S.J. Sharkis, A low level of reactive oxygen species selects for primitive hematopoietic stem cells that may reside in the low-oxygenic niche. Blood **110**, 3056–3063 (2007)

23. A.-R. Ji, S.-Y. Ku, M.S. Cho, Y.Y. Kim, Y.J. Kim, S.K. Oh, S.H. Kim, S.Y. Moon, Y.M. Choi, Reactive oxygen species enhance differentiation of human embryonic stem cells into mesendodermal lineage. Exp. Mol. Med. **42**, 175–186 (2010)
24. A. Gomes, N. Price, A. Ling, J. Moslehi, M.K. Montgomery, L. Rajman, J. White, J. Teodoro, C. Wrann, B. Hubbard, E. Mercken, C. Palmeira, R. deCabo, A. Rolo, N. Turner, E. Bell, D. Sinclair, Declining NAD+ induces a pseudohypoxic state disrupting nuclear-mitochondrial communication during aging. Cell **155**, 1624–1638 (2016)
25. P. Rimmelé, C. Bigarella, R. Liang, B. Izac, R. Dieguez-Gonzalez, G. Barbet, M. Donovan, C. Brugnara, J. Blander, D. Sinclair, S. Ghaffari, Aging-like phenotype and defective lineage specification in SIRT1-deleted hematopoietic stem and progenitor cells. Stem Cell Rep. **3**, 44–59 (2016)
26. C.D. Vulpe, S. Packman, Cellular copper transport. Annu. Rev. Nutr. **15**, 293–322 (1995)
27. E.D. Harris, Copper transport: an overview, in *Proceedings of the Society for Experimental Biology and Medicine*, vol. 196 (Society for Experimental Biology and Medicine, New York, N.Y., 1991), pp. 130–140
28. J.R. Prohaska, Role of copper transporters in copper homeostasis. Am. J. Clin. Nutr. **88**, 826S–829S (2008)
29. A. Zabek-Adamska, R. Drozdz, J.W. Naskalski, Dynamics of reactive oxygen species generation in the presence of copper(II)-histidine complex and cysteine. Acta Biochim. Pol. **60**, 565–571 (2013)
30. T.S. Koskenkorva-Frank, G. Weiss, W.H. Koppenol, S. Burckhardt, The complex interplay of iron metabolism, reactive oxygen species, and reactive nitrogen species: Insights into the potential of various iron therapies to induce oxidative and nitrosative stress. Free Radic. Biol. Med. **65**, 1174–1194 (2013)
31. B. Zhou, J. Gitschier, hCTR1: a human gene for copper uptake identified by complementation in yeast. Proc. Natl. Acad. Sci. U.S.A. **94**, 7481–7486 (1997)
32. L.B. Moller, C. Petersen, C. Lund, N. Horn, Characterization of the hCTR1 gene: genomic organization, functional expression, and identification of a highly homologous processed gene. Gene **257**, 13–22 (2000)
33. Y.M. Kuo, B. Zhou, D. Cosco, J. Gitschier, The copper transporter CTR1 provides an essential function in mammalian embryonic development. Proc. Natl. Acad. Sci. U.S.A. **98**, 6836–6841 (2001)
34. T. Haremaki, D.C. Weinstein, Xmc mediates Xctr1-independent morphogenesis in Xenopus laevis. Dev. Dyn.: An official publication of the American Association of Anatomists **238**, 2382–2387 (2009)
35. H.-Q. Duong, Y.B. Hong, J.S. Kim, H.-S. Lee, Y.W. Yi, Y.J. Kim, A. Wang, W. Zhao, C.H. Cho, Y.-S. Seong, I. Bae, Inhibition of checkpoint kinase 2 (CHK2) enhances sensitivity of pancreatic adenocarcinoma cells to gemcitabine. J. Cell Mol. Med. **17**, 1261–1270 (2013)
36. E.K.-H. Han, T. Mcgonigal, C. Butler, V.L. Giranda, Y. Luo, Characterization of Akt overexpression in MiaPaCa-2 cells: prohibitin is an Akt substrate both in vitro and in cells. Anticancer Res. **28**, 957–963 (2008)
37. Y. Iwagami, H. Eguchi, H. Nagano, H. Akita, N. Hama, H. Wada, K. Kawamoto, S. Kobayashi, A. Tomokuni, Y. Tomimaru, M. Mori, Y. Doki, miR-320c regulates gemcitabine-resistance in pancreatic cancer via SMARCC1. Br. J. Cancer **109**, 502–511 (2013)
38. A.A. Yunis, G.K. Arimura, D.J. Russin, Human pancreatic carcinoma (MIA PaCa-2) in continuous culture: sensitivity to asparaginase. Int. J. Cancer. Journal international du cancer **19**, 128–135 (1977)
39. I.S. Ahmed, H.J. Rohe, K.E. Twist, M.N. Mattingly, R.J. Craven, Progesterone receptor membrane component 1 (Pgrmc1): a heme-1 domain protein that promotes tumorigenesis and is inhibited by a small molecule. J. Pharmacol. Exp. Therapeut. **333**, 564–573 (2010)
40. R.M. Losel, D. Besong, J.J. Peluso, M. Wehling, Progesterone receptor membrane component 1-many tasks for a versatile protein. Steroids **73**, 929–934 (2008)
41. H.J. Rohe, I.S. Ahmed, K.E. Twist, R.J. Craven, PGRMC1 (progesterone receptor membrane component 1): a targetable protein with multiple functions in steroid signaling, P450 activation and drug binding. Pharmacol. Therapeut. **121**, 14–19 (2009)

42. A.M. Friel, L. Zhang, C.A. Pru, N.C. Clark, M.L. McCallum, L.J. Blok, T. Shioda, J.J. Peluso, B.R. Rueda, J.K. Pru, Progesterone receptor membrane component 1 deficiency attenuates growth while promoting chemosensitivity of human endometrial xenograft tumors. Cancer Lett. **356**, 434–442 (2015)

43. X. Zhu, Y. Han, Z. Fang, W. Wu, M. Ji, F. Teng, W. Zhu, X. Yang, X. Jia, C. Zhang, Progesterone protects ovarian cancer cells from cisplatin-induced inhibitory effects through progesterone receptor membrane component 1/2 as well as AKT signaling. Oncol. Rep. **30**, 2488–2494 (2013)

44. M.A. Cahill, Progesterone receptor membrane component 1: an integrative review. J. Steroid Biochem. Mol. Biol. **105**, 16–36 (2007)

45. J.J. Peluso, A. Pappalardo, R. Losel, M. Wehling, Progesterone membrane receptor component 1 expression in the immature rat ovary and its role in mediating progesterone's antiapoptotic action. Endocrinology **147**, 3133–3140 (2006)

46. H. Neubauer, S.E. Clare, W. Wozny, G.P. Schwall, S. Poznanovic, W. Stegmann, U. Vogel, K. Sotlar, D. Wallwiener, R. Kurek, T. Fehm, M.A. Cahill, Breast cancer proteomics reveals correlation between estrogen receptor status and differential phosphorylation of PGRMC1. Breast Cancer Res.: BCR **10**, R85 (2008)

47. P.P. Adhikary, Role of Progesterone Receptor Membrane Component 1 (PGRMC1) in Cancer Cell Biology. Ph.D. thesis, Charles Sturt University, 2016

48. E.H. Sarsour, M.G. Kumar, L. Chaudhuri, A.L. Kalen, P.C. Goswami, Redox control of the cell cycle in health and disease. Antioxid. Redox Signal. **11**, 2985–3011 (2009)

49. M. Brand, D. Nicholls, Assessing mitochondrial dysfunction in cells. Biochem. J. **435**, 297–312 (2011)

50. The Official Journal of the Society for, N. Yadava, D.G. Nicholls, Spare respiratory capacity rather than oxidative stress regulates glutamate excitotoxicity after partial respiratory inhibition of mitochondrial complex I with rotenone. J. Neurosci Neuroscience **27**, 7310–7317 (2007)

51. L.A. Kunz-Schughart, Multicellular tumor spheroids: intermediates between monolayer culture and in vivo tumor. Cell Biol. Int. **23**, 157–161 (1999)

52. G. Hamilton, Multicellular spheroids as an in vitro tumor model. Cancer Lett. **131**, 29–34 (1998)

53. F. Hirschhaeuser, H. Menne, C. Dittfeld, J. West, W. Mueller-Klieser, L.A. Kunz-Schughart, Multicellular tumor spheroids: an underestimated tool is catching up again. J. Biotechnol. **148**, 3–15 (2010)

54. W. Mueller-Klieser, Multicellular spheroids. A review on cellular aggregates in cancer research. J. Cancer Res. Clin. Oncol. **113**, 101–122 (1987)

55. D. Shweiki, M. Neeman, A. Itin, E. Keshet, Induction of vascular endothelial growth factor expression by hypoxia and by glucose deficiency in multicell spheroids: implications for tumor angiogenesis. Proc. Natl. Acad. Sci. **92**, 768–772 (1995)

56. K.R. Frenzel, R.M. Saller, J. Kummermehr, S. Schultz-Hector, Quantitative distinction of cisplatin-sensitive and -resistant mouse fibrosarcoma cells grown in multicell tumor spheroids. Cancer Res. **55**, 386–391 (1995)

57. K.A. Krohn, J.M. Link, R.P. Mason, Molecular Imaging of Hypoxia. J. Nucl. Med. **49**, 129S–148S (2008)

58. R.M. Sutherland, B. Sordat, J. Bamat, H. Gabbert, B. Bourrã, Oxygã nationand Differentiation in Multicellular Colon Carcinoma1 Spheroids of Human. Regulation, 5320–5329 (1986)

59. U. Berchner-Pfannschmidt, S. Frede, C. Wotzlaw, J. Fandrey, Imaging of the hypoxia-inducible factor pathway: insights into oxygen sensing. Eur. Respir. J. **32**, 210–217 (2008)

60. A.K. Renfrew, N.S. Bryce, T.W. Hambley, Delivery and release of curcumin by a hypoxia-activated cobalt chaperone: a XANES and FLIM study. Chem. Sci. **4**, 3731–3739 (2013)

61. R.H. Thomlinson, L.H. Gray, The histological structure of some human lung cancers and the possible implications for radiotherapy. Br. J. Cancer **9**, 539–549 (1955)

62. S. Zhang, M. Hosaka, T. Yoshihara, K. Negishi, Y. Iida, S. Tobita, T. Takeuchi, Phosphorescent light-emitting iridium complexes serve as a hypoxia-sensing probe for tumor imaging in living animals. Cancer Res. **70**, 4490–4498 (2010)

Chapter 7
Ex Vivo Studies

As outlined in Chap. 6, cultured cells have become an indispensable technology in various branches of life sciences. However, there are some concerns associated with the study of cultured cells. In a living organism, cells exist in a three-dimensional geometry whereas under laboratory conditions, cells are grown on a two-dimensional substrate. As a result, the cell-to-cell interactions that may be crucial for normal cellular functions are lost [1]. Also, cells naturally grow in a heterogeneous environment (in presence of other cell types), rather than only one cell type as for cell culture. It has been observed that over a period of continuous growth and sub-culturing in laboratory conditions the morphological and biochemical characteristics of the cells deviate from those originally found in the donor animals [2–4], so they are no longer a good physiological model. Such changes have been associated with genetic instability of cultured cells, which ultimately leads to heterogeneity in cells [5]. Therefore, there has recently been greater interest in performing ex vivo investigations using cells isolated directly from animals, termed primary cells [6]. These primary cells offer the similar ease of handling and imaging cultured cells whilst possessing unperturbed biochemical and physiological state [6, 7]. However, isolation of primary cells is more arduous. Herein, the developed fluorescent redox probes have been employed to perform investigations on the oxidative capacity of primary cells isolated from mice (Fig. 7.1). Hepatocytes (liver cells) and haematopoietic (blood-forming cells) have been interrogated.

Parts of the text and figures of this chapter are reprinted with permission from *Antioxidants and Redox Signalling*, Volume 24, Issue 13, published by Mary Ann Leibert, Inc., New Rochelle, NY., and Organic and Biomolecular Chemistry, Issue 24, with permission from the Royal Society of Chemistry.

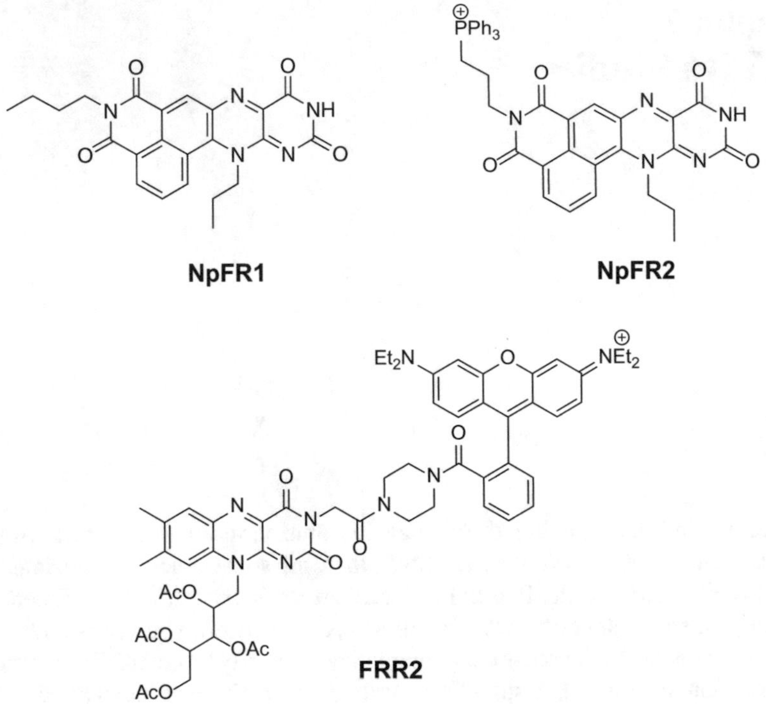

Fig. 7.1 Chemical structures of the redox probes utilised in ex vivo experiments

7.1 Hepatocytes

The liver plays a crucial role in maintaining the body's metabolic balance and a damaged liver can result in a variety of diseases and metabolic syndromes. The liver is the chief detoxifying organ in a mammalian body and comprises 70–85% hepatocytes, which perform most of the organ's functions, including metabolism of carbohydrates and fats, and synthesis and storage of proteins. Two crucial functions of hepatocytes are detoxification and excretion of endogenous metabolites and xenobiotics, such as drugs, pesticides, preservatives, pollutants and toxins [8, 9]. Isolated primary hepatocytes therefore serve as a pragmatic ex vivo model to investigate many aspects of liver physio-pathology, particularly to explore mechanisms of detoxification and drug-metabolism [10]. In addition, hepatocytes provide a unique model to investigate the influence of external stress on the liver, and to screen potential stress.

The substantial capacity of hepatocytes to effectively metabolise and detoxify results in the production of significantly higher levels of ROS than in other mammalian cells. Protection against high oxidant levels is made possible by the multi-layer antioxidant defence systems that function in hepatocytes [11]. These include low-molecular weight antioxidants such as glutathione (2–10 mM), vitamin C (40–140 μM) and vitamin E (10–40 μM), and antioxidant enzymes such as catalase, glutathione peroxidase and superoxide dismutases, which effectively counteract the elevation in ROS production [10].

As in any other cell, redox regulation plays a vital role in maintaining health and functions of hepatocytes. In fact, impaired antioxidant defence systems and the resulting oxidative stress, have been implicated in most chronic liver diseases— alcoholic liver disease (ALD) [12], non-alcoholic fatty liver disease (NAFLD) [13], hepatic encephalopathy (HE) [14], liver fibroproliferative diseases [15] and Hepatitis C infection [16]. While several studies have gathered biochemical evidence to confirm this, the intricacies of the underlying mechanisms are yet to be deciphered [17].

7.1.1 Oxidative Stress in Primary Mouse Hepatocytes

Considering the vital role hepatocytes play in lipid metabolism, these cells provide an ideal system to investigate the protective role of lipids against elevated ROS levels. Isolated primary mouse hepatocytes treated with **NpFR1** were interrogated to determine the extent of oxidative stress caused by hydrogen peroxide, and the protective roles of lipids. Dr Moumita Paul and Ms Shilpa Nagarajan isolated and prepared the primary mouse hepatocytes. $11,11$-D_2-Linoleic acid and $11,11,14,14$–D_4-α–linolenic acid were provided by Dr Shchepinov.

The isolated hepatocytes were plated at a density of 10,000 cells/well in a 96-well plate in low glucose DMEM supplemented and incubated overnight. In order to determine the concentration of peroxide capable of perturbing the redox homoeostasis of hepatocytes, cells were treated with incremental concentrations of H_2O_2 ranging from 100–500 μM for 30 min. The cells were then washed and incubated in low glucose DMEM containing 20 μM **NpFR1** for a period of 15 min. This was followed by washing the hepatocytes twice with PBS, and maintaining them in FACS buffer (PBS supplemented with 0.1% FBS) for the duration of imaging. The images were acquired using BD Pathway™ 855, a high-throughput live cell imaging station, upon excitation with a 405 nm laser. **NpFR1** emission was recorded using a 450 nm long pass filter, with a higher intensity of fluorescence indicating a higher oxidative capacity. Image analysis was performed using FIJI (National Institute of Health).

As shown in Fig. 7.2, H_2O_2 concentrations below 400 μM did not show any significant increase in oxidative capacity compared to the untreated cells, suggesting that the antioxidant defences of hepatocytes were able to cope with this level of oxidative insult.

However, hepatocytes treated with 400 μM H_2O_2 show a sudden 3-fold increase in fluorescence, demonstrating that, at this concentration of H_2O_2, the antioxidant defence mechanisms of hepatocytes fail, and are unable to restore the redox homoeostasis. These results are consistent with findings that induction of necrosis occurs in primary hepatocytes upon treatment with H_2O_2 concentrations between 300–400 μM [18].

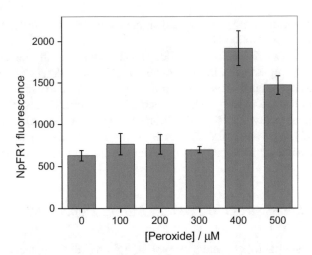

Fig. 7.2 The **NpFR1** fluorescence emission from hepatocytes treated with a range of H_2O_2 concentrations. Bars represent the average emission intensity ($\lambda_{ex} = 405$ nm and $\lambda_{em} = 450$ nm long pass). Error bars represent standard deviation ($n = 3$)

7.1.2 Oxidation of Polyunsaturated Fatty Acids (PUFAs)

One of the numerous downstream effects of oxidative stress is lipid peroxidation, and considering the role of hepatocytes in lipid metabolism, it is unsurprising that lipid peroxidation has been observed in most chronic hepatic diseases. Polyunsaturated fatty acids (PUFAs) such as linoleic and α-linolenic acid are essential fatty acids which act as precursors for the production of higher PUFAs [19]. Owing to their pivotal role in lipid membranes, PUFAs are highly susceptible to ROS-mediated oxidation. The oxidation of PUFAs by ROS is a chain reaction initiated by a small number of ROS, usually the hydroxyl radical ($^{\bullet}$OH), with subsequent oxidation of large numbers of PUFA molecules, and generation of an array of lipid oxidation products such as acrolein, 4-hydroxynonenal and malionaldehyde (Fig. 7.3). These toxic lipid oxidation products result in changes in membrane fluidity and downstream signalling of pro-fibrotic factors (leading to fibrosis) [8]. Furthermore, in the ROS-mediated oxidation of PUFAs, the rate-limiting step is the removal of hydrogen from the bis-allylic site (Fig. 7.3).

Dr Mikhail S. Shchepinov (Retrotope, Inc.) reported that, in mammalian models of Friedreich's ataxia, PUFAs with deuterated bis-allylic sites (11,1-D_2-linoleic acid and 11, 11, 14, 14–D_4-α-linolenic acid, Fig. 7.4) demonstrate exceptional protection against treatment with ferric ammonium citrate (FAC, an iron source) and buthionine sulfoximine (BSO, an inhibitor of glutathione synthesis), both of which are known to elevate cellular ROS levels [20]. Furthermore, the toxic effects of FAC and BSO were exacerbated by the non-deuterated counterparts of these PUFAs (H-linoleic acid and H-α-linolenic acid). The results from their study indicated the potential of deuterated-PUFAs as effective therapy for Friedreich's ataxia.

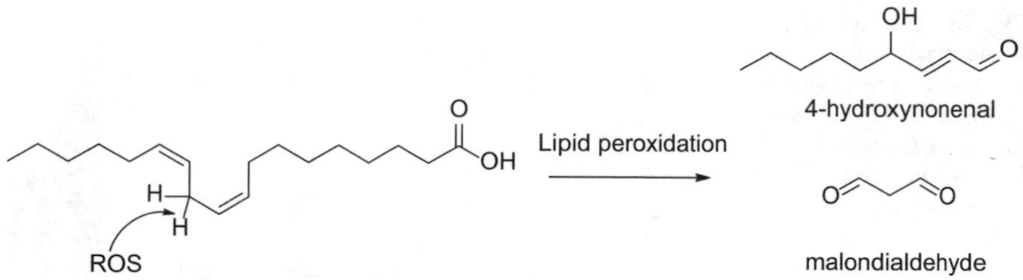

Fig. 7.3 ROS-mediated lipid peroxidation results in the formation of toxic carbonyl compounds-4-hydroxynonenal and malondialdehyde. Rate driven step is the hydrogen abstraction from the bis-allylic site

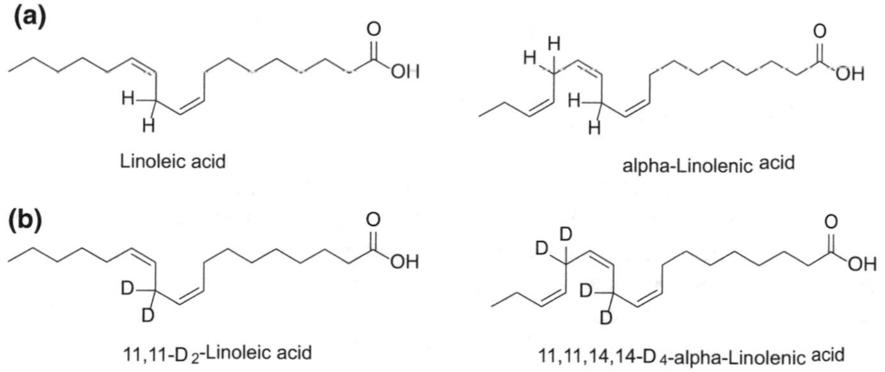

Fig. 7.4 Chemical structures of essential PUFAs. **a** linoleic and alpha-linolenic acid, **b** deuterated counterparts of linoleic and alpha-linolenic acid where the hydrogen atoms at the bis-allylic sites have been isotopically protected

7.1.3 Assessing the Antioxidant Properties of PUFAs

The antioxidant properties of palmitate and deuterated PUFAs were investigated in isolated primary mouse hepatocytes. This was accomplished by seeding 10,000 hepatocytes in each well of a 96 well plate overnight followed by an overnight treatment with palmitate and deuterated linoleic and linolenic acids. Following chronic lipid treatments, hepatocytes were treated with H_2O_2 (100–400 μM) for 30 min. Hepatocytes were then washed and incubated with 20 μM **NpFR1** for 15 min, then washed and imaged in FACS buffer ($\lambda_{ex} = 405$ nm and $\lambda_{em} = 450$ nm long pass). Figure 7.5 demonstrates that all three lipids tested exhibit some degree of antioxidant protection against the 400 μM H_2O_2-insult. Deuterated linolenic acid was observed to be more effective at combating oxidative stress than linoleic acid, which is in accordance with the results obtained by Shchepinov and co-workers [19, 21]. The results confirm that both palmitate and the deuterated linolenic acid show promising antioxidant properties in hepatocytes.

Fig. 7.5 The **NpFR1**
fluorescence emission from
hepatocytes pre-treated with
palmitate and dueterated
linoleic and linolenic acids
followed by treatment with a
range of H_2O_2
concentrations. Bars
represent the average
emission intensity
($\lambda_{ex} = 405$ nm and
$\lambda_{em} = 450$ nm long pass).
Error bars represent standard
deviation ($n = 3$)

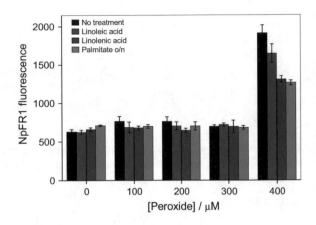

7.1.4   Conclusions

These experiments demonstrate the potential of **NpFR1** as a reporter of oxidative
stress levels in hepatocytes. The fluorescent response of **NpFR1** suggests that the
antioxidant defences of hepatocytes are capable of protecting the cells against a H_2O_2
insult of up to $300\,\mu M$, beyond which the oxidative capacity of cells increase dra-
matically. Furthermore, deuterated linolenic acid and palmitate exhibited significant
protection against H_2O_2-mediated oxidative stress.

7.2   Haematopoietic Cells

Haematopoietic stem cells (HSCs) that reside in the bone marrow possess a unique
ability to differentiate into a variety of mature cells that constitute the blood and
lymph—two cardinal fluids of the vertebrates's circulatory system [22]. During the
process of haematopoiesis, the HSCs proliferate via asymmetric division, which
results in the formation of two distinct daughter cells (Fig. 7.6) [22]. One of the
daughter cells is completely identical to the original stem cell, so the pool of stem
cells is never depleted, and the second daughter cell is programmed to differentiate
into one of two cell lineages—the myeloid or the lymphoid [22, 23].

The myeloid progenitor cell differentiates to produce megakaryocytes (platelet
forming cells), erythrocytes (red blood cells), mast cells (a type of white blood cell)
and myeloblasts (Fig. 7.6) [23, 24]. The myeloblasts further differentiate into more
specialised white blood cells—the eosinophils, basophils, neutrophils and monocytes
[25, 26]. These monocytes are the precursors for the generation of macrophages.
While eosinophils and basophils modulate inflammatory responses, neutrophils and
macrophages are phagocytes responsible for destroying pathogens. On the other
hand the lymphoid progenitor cell differentiates into two sub-types of white blood

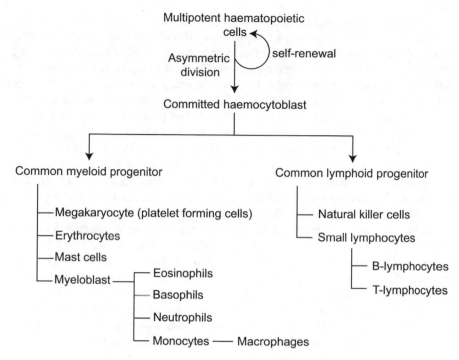

Fig. 7.6 An outline of the process of haematopoiesis

cells responsible for mediating vertebrates' immunological defences; the natural killer cells (NK cells), and the lymphocytes (Fig. 7.6) [25, 26]. The NK cells are a component of the non-specific, or innate, immunity that is responsible for destroying compromised host cells, such as cancerous cells or virus-infected cells, whereas the lymphocytes are major cellular agents of adaptive immunity [24, 25]. Following their formation, some lymphocytes undergo maturation within the bone marrow to form B-lymphocytes involved in humoral immunity (antibody-mediated immunity), while the rest travel to the thymus gland (a secondary lymphoid organ) for maturation into T-lymphocytes, which play a central role in the cell-mediated immunity. Besides thymus, the other crucial secondary lymphoid organ is the spleen, which is the site of immune cell function, and destruction of aged and damaged red blood cells [22].

7.2.1   Redox State of Haematopoietic Cells

The process of haematopoiesis results in the formation of a variety of cell types, each having its own specialised functions and unique metabolic profiles. Mitochondrial ROS levels play a critical role in the differentiation of haematopoietic stem cells [27]. Terminally differentiated erythrocytes do not have mitochondria, and therefore must proceed through mitophagy (selective degradation of mitochondria) to become fully functional [28]. Mitophagy in erythrocytes is regulated by the Bcl-2 family member Nix [28]. Mice lacking both copies fail to clear mitochondria from their maturing reticulocytes, and therefore develop reticulocytosis, leading to fatal anaemia [29, 30]. The aim of this section of the work was to investigate the variations in oxidative capacity of various haematopoietic cells isolated from mouse bone marrow, thymus and spleen. Isolated cells were treated with **NpFR2** and assessed using a flow cytometer. Isolation of haematopoietic cells from mice was performed by Dr Stuart Fraser at the Discipline of Physiology, School of Medicine, University of Sydney.

While flow cytometers are commonly used to record the fluorescence emitted from a single cell, there is a variety of valuable information that can be obtained as a result of a laser shining on an individual cell. Besides fluorescence, two important components that contribute to cell identification are forward scatter and side scatter, which measure the light scattered by the cells as they flow through the flow cytometer (Fig. 7.7) [31]. Forward scatter refers to the light scattered in the direction of the laser beam, and is measured by an detector on the opposite side of the cell. For larger cells, more light is scattered, so a higher forward scatter will be recorded than for the smaller cells. Side scatter refers to the light that is scattered at different angles from the laser path after it strikes the cell. Cells with a higher cytoplasmic complexity, usually due to the presence of organelles or granules, scatter larger amounts of light (Fig. 7.7) [31]. Combining the information of forward and side scatter can therefore give a profile of cell size and granularity [31].

The bone marrow is a heterogeneous mixture of a variety of cell types which vary in both size and granularity. As a result, each cell type has a distinct and well-defined profile on the forward versus side scatter plot. Haematopoietic cells isolated from mouse bone marrow by Dr Stuart Fraser were screened for their forward and side scatter profiles. As shown in Fig. 7.8, larger and more granular macrophages produce a population with high forward and side scatter. Lymphoblasts and myeloblasts are smaller than macrophages, and therefore have a lower forward scatter, but myeloblasts, being more granular, produce a separate population with higher side scatter. Smaller agranular erythroblasts and erythrocytes form a popu-

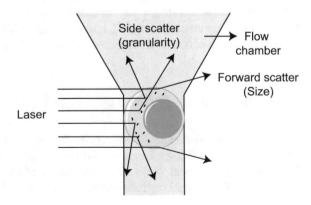

Fig. 7.7 Light scattered in different directions as the laser interrogates the cell. The direction of light scattered in forward and side direction correspond to cell size and granularity respectively

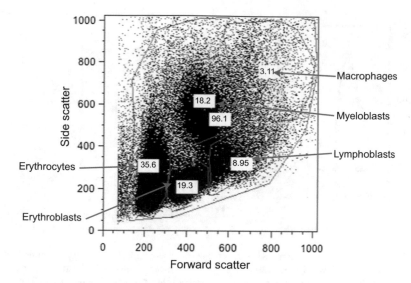

Fig. 7.8 A dot plot of forward versus side scatter showing different types of haematopoietic cells present in the mouse bone marrow. Numbers indicate relative percentage of live cells (red) and total cells corresponding to the indicated population (blue). Reprinted from Organic and Biomolecular Chemistry, Issue 24, with permission from the Royal Society of Chemistry

lation with lower side scatter (Fig. 7.8). Erythroblasts, the nucleated precursors of erythrocytes, are larger than the latter, and therefore produce a separate population with higher forward scatter.

To gain an insight into the mitochondrial oxidative capacity of the haematopoietic cells in the bone marrow, the isolated cells were treated **NpFR2** (20 μM, 15 min) and then assessed for fluorescence using a flow cytometer. **NpFR2** fluorescence could be detected in fluorescent channels 1 (488 nm excitation, 530/30 nm emission) and 2 (488 nm excitation, 585/42 nm emission) of the BD FACScan flow cytometer. Bone marrow cells showed 4 different populations with range of **NpFR2** fluorescent signals from negative to very bright (Fig. 7.9). Each population was gated and their forward and side scatter profiles examined to identify cells. The forward and side scatter profile of the most fluorescent population D, comprising of 45% of the bone marrow cells, consistent with that of macrophages (Fig. 7.9). The forward and side scatter profile of populations A indicated a major population of erythrocytes. Populations B and C represented a combination of erythrocytes and lymphoblasts, and mostly erythroblasts respectively.

In order to specifically identify the populations to which these cells belonged, bone marrow cells were incubated with various antibodies conjugated to fluorophores (Table 7.1) that exhibit far-red fluorescence (channel 4, 633 nm excitation, 670 long pass emission filter).

Erythroid cells (that form red blood cells) were identified by use of the lineage-specific molecule Ter-119. Two distinct populations could be detected; one with

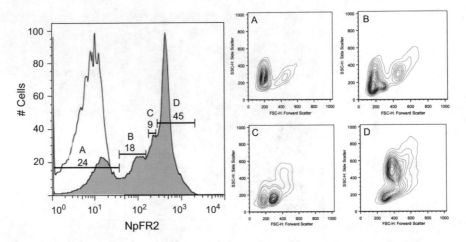

Fig. 7.9 Detection of mitochondrial ROS species in haematopoietic cell types using **NpFR2** (20 μM, 15 min) by flow cytometry. Histograms of bone marrow cells (left) showing 4 populations with different **NpFR2** fluorescence intensities (tinted profile) compared to unstained cells (unfilled profile). Numbers indicate frequency of total live, gated events. Contour plots (right) showing relative frequency of the populations for forward versus side scatter of different populations. Reprinted from Organic and Biomolecular Chemistry, Issue 24, with permission from the Royal Society of Chemistry

Table 7.1 Antibodies used to identify distinct haematopoietic cell types

Cell type	Antigen	Clone	Fluorochrome	Commercial source
All haematopoietic cells	CD45	30-F11	Allophycocyanin	eBioscience
Erythroid cells	Ter-119	Ter-119	Alexa Fluor 647	BioLegend
Megakaryocytes	CD41	eBioMWReg30	Allophycocyanin	eBioscience
Macrophages	F4/80	BM8	eFluor660	eBioscience
Mast cells	Allergen	TX83	eFluor660	eBioscience
T-lymphocytes	CD4	GK1.5	Allophycocyanin	eBioscience

Ter-119 expression and high levels of **NpFR2** fluorescence, representing developing erythroid cells, and a second with Ter-119 expression, but no **NpFR2** fluorescence. This second population corresponds to the phenotype of mature erythrocytes, which express Ter-119 but lack mitochondria (Fig. 7.10).

Macrophages can be identified by the surface expression of the F4/80 antibody [26]. Cells identified as macrophages exhibited two populations having medium and high levels of **NpFR2** fluorescence. This suggests that bone marrow macrophages have medium and high levels of mitochondrial ROS. A similar pattern of **NpFR2** fluorescence was observed in CD41-expressing megakaryocytes (which form platelets). Mast cells, identified by the expression of allergen, exhibited relatively uniform **NpFR2** fluorescence. The anti-CD45 antibody conjugated to allophycocyanin was

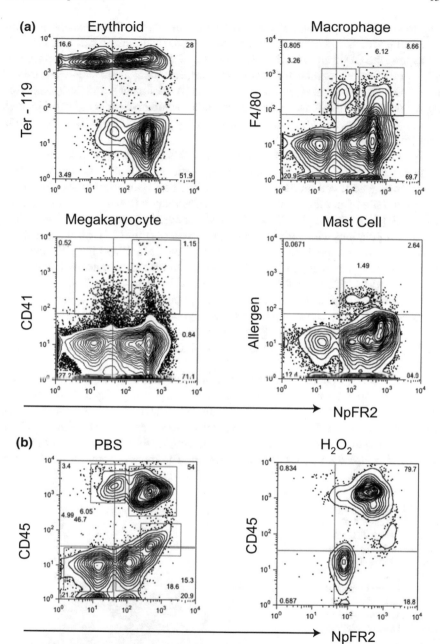

Fig. 7.10 a Contour plots showing **NpFR2** fluorescence in various bone marrow-derived haematopoietic cell types: Ter-119 positive (erythroid cells); F4/80-expressing (macrophages); CD41-expressing (megakaryocytes) and allergen-expressing (mast cells). **b** Contour plots showing **NpFR2** fluorescence in CD45-labelled haematopoietic cells after treatment with vehicle control (PBS) and H_2O_2. Reprinted from Organic and Biomolecular Chemistry, Issue 24, with permission from the Royal Society of Chemistry

used to identify all haematopoietic cells (Population A, Fig. 7.10), except maturing erythroid or red blood cells, which are the only haematopoietic cell types that do not express the CD45 antigen (Population B, Fig. 7.10). Furthermore, to examine the effect of exogenous oxidative stress conditions, haematopoietic cells were treated with H_2O_2. Compared to PBS-treated control cells, treatment with hydrogen peroxide demonstrated an increase in their mitochondrial oxidative stress indicated by high **NpFR2** fluorescence in all bone marrow cells, regardless of their CD45 expression level (Fig. 7.10).

NpFR2 was also employed to report on the oxidative capacity of mitochondria in the same lineage of cells in different haematopoietic organs; the bone marrow (site of most hematopoietic cell production), thymus (site of T lymphocyte maturation) and the spleen (site of immune cell function and destruction of aged and damaged red blood cells) (Figs. 7.11 and 7.12). As discussed above, the bone marrow showed four different populations with varied **NpFR2** fluorescence (Fig. 7.11). The cells isolated from the thymus resolved into two populations—one exhibiting high mitochondrial ROS (**NpFR2**-bright peak) and a smaller population with comparatively lower levels of mitochondrial ROS (**NpFR2**-medium peak) in the thymus (Fig. 7.11).

The spleen cells showed two different cell populations, one of which exhibited bright **NpFR2** fluorescence suggesting that high mitochondrial oxidative capacity could possibly represent the highly active immune cells. The other population, with very low **NpFR2** fluorescence (Fig. 7.11), is likely to correspond to the aged and damaged erythrocytes.

After obtaining a broad perspective of the mitochondrial redox state of cells in different haematopoietic organs, the mitochondrial oxidative capacity of T-lymphocytes residing in each of these organs was evaluated. The isolated cells were treated with a mixture of **NpFR2** and allophycocyanin-conjugated CD4 antibody. CD4-expressing T-lymphocytes arise from rare progenitors in the bone marrow and showed low/negative **NpFR2** fluorescence (Fig. 7.12a). The thymus is the primary site of

Fig. 7.11 Offset histograms of **NpFR2** fluorescent signal, showing different populations in thymus, bone marrow (BM) and spleen single cell suspensions. Reprinted from Organic and Biomolecular Chemistry, Issue 24, with permission from the Royal Society of Chemistry

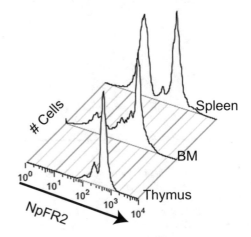

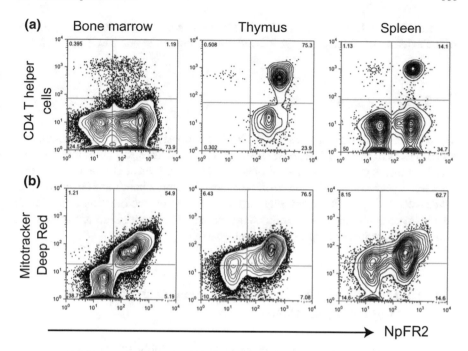

Fig. 7.12 **a** NpFR2 fluorescence in CD4-expressing T-helper lymphocytes from different organs. Mitochondrial ROS as detected by **NpFR2** varies according to the site the CD4-expressing cell is found. **b** Comparison of mitochondrial number (Mitotracker) and mitochondrial ROS (**NpFR2**) in different haematopoietic organs. Data shows representative flow cytometry profiles from at least three independent experiments. Reprinted from Organic and Biomolecular Chemistry, Issue 24, with permission from the Royal Society of Chemistry

CD4-positive cell expansion and education. **NpFR2** fluorescence is observed in the vast majority of these cells, reflecting their highly proliferative state. In the spleen, mature functional CD4-positive T-helper lymphocytes respond to immune challenges. This may be reflected by their high **NpFR2**-fluorescence (Fig. 7.12a). A small population (~1%) of CD4-positive cells lacking **NpFR2** fluorescence were observed. These may correspond to memory CD4 T-helper lymphocytes that reside in the spleen following exposure to pathogens in a highly quiescent state (Fig. 7.12a).

Finally, to confirm that the changes observed in **NpFR2** fluorescence reflected the changes in mitochondrial redox status rather than mitochondrial number, cells isolated from different haematopoietic organs were co-incubated with **NpFR2** and Mitotracker DeepRed FM (Fig. 7.12b). Bone marrow cells can be grouped into two cell types: Mitotracker DeepRed-negative, **NpFR2**-negative (which most likely represent mature erythrocytes lacking mitochondria), and a population of Mitotracker DeepRed, **NpFR2**-double positive cells showing fluorescence from both. Most cells in the thymus showed fluorescence from both dyes (Fig. 7.12b). A similar observation could be made for spleen samples, but a population that showed Mitotracker Deep

Red fluorescence in the absence of NpFR2 fluorescence indicative of a cell type with mitochondria that did not contain ROS was not observed in any case (Fig. 7.12b).

7.2.2   Conclusions

These experiments demonstrate that **NpFR2** is a robust sensor of the mitochondrial redox state in haematopoietic cells. **NpFR2** also shows immense potential as a sensor for detecting the varying levels of mitochondrial ROS and can be combined with other fluorescent probes and antibodies to dissect the role of mitochondrial ROS in different blood cells during development, maturation, proliferation and function. **NpFR2** can be employed to provide unique information about mitochondrial redox state beyond cultured cells in more complex biological systems.

7.3   Imaging Oxidative Capacity of Haematopoietic Cells During Embryogenesis

Mammalian embryogenesis takes place in utero, a highly hypoxic environment, therefore hypoxia-inducible factors are critical to the regulation of embryonic haematopoiesis [27]. Erythropoiesis (production of red blood cells) in mammalian embryos and adults occurs in several waves, resulting in the production of at least 5 distinct oxygen-transporting erythroid cell types [32]. In mice, the early phase of embryogenesis is marked by the formation of three germ layers—ectoderm, mesoderm and endoderm. The first progenitors for primitive erythroid cells appear approximately 7.5 *days post coitum* (dpc, referring to days after copulation) and by 8.5 dpc, the progenitors have developed into the primitive erythroid (EryP) lineage. This lineage comprises of red blood cells that are around 6-fold larger than their adult equivalents, which dominate until 14.5 dpc and then almost disappear by birth [33, 34]. Upon entering the circulation, a series of synchronised morphological changes begins to unfold in the EryP cells, including reduction in cytoplasmic content and cell size [33, 35]. Around 12.5–13.5 dpc the nucleus disappears and all EryP cells cease dividing. 13.5–15.5 dpc, there is a pronounced increase in the numbers of enucleated EryP cells [33, 35].

The foetal liver begins to operate as a haematopoietic organ around 12.5 dpc, and is involved in the rapid generation of adult-type definitive erythroid (EryD) cells. The production of these small and anuclear EryD cells peaks at 14.5 dpc, at which point they begin to outnumber EryP in the bloodstream [32, 36]. Shortly before birth, the bone marrow becomes the predominant site of haematopoiesis. These erythrocytes are slightly smaller than those produced by the fetal liver. Stresses such as chronic erythroid diseases (β-thalassemia or sickle cell anaemia) or heavy blood loss result in an auxiliary wave of erythropoiesis, which is characterised by the transfer of immature erythroblasts from the bone marrow to the spleen [37].

During embryonic erythropoiesis, rapid variations have been observed in the expression levels of genes for hypoxia response and for the control of mitochondrial function and mitophagy [38]. Cellular redox homeostasis, particularly in the mitochondria, has been reported to be involved in regulating the differentiation and maturation processes that occur during embryonic and adult haematopoiesis [39–41]. The aim of this section of the work was to monitor the variations in mitochondrial ROS in cells of a similar biological origin and function. To this end, **FRR2** fluorescence was measured in developing embryonic mouse erythroid (red blood) cells, which are in a hypoxic environment compared to the adult mouse erythroid cells generated in the adult bone marrow.

Single cell suspensions of embryonic blood, foetal liver, adult blood, adult bone marrow and adult spleen at 9.5, 11.5, 12.5, 14.5 and 15.5 dpc were incubated with a combination of **FRR2** (20 μM) and Ter-119 (antibody marker for erythroid cells) for 15 min, and then assessed for fluorescence using a flow cytometer. Analysis of embryonic blood 9.5 dpc treated with **FRR2** (Fig. 7.13) showed a single population with uniform **FRR2** fluorescence corresponding to the circulating primitive erythroid cells, which contain a nucleus and metabolically active mitochondria. By 11.5–12.5 dpc, two distinct populations (labelled 1 and 2) can be seen in the circulating blood, as identified by their **FRR2** fluorescence profiles (Fig. 7.13). Population 1 shows higher **FRR2**-red fluorescence than population 2. When the size and granularity of the former population were assessed, cells were found to comprise a population of larger cells (the yolk sac-derived definitive erythroblasts), and a population of smaller cells (the matured, erythrocytes that lack a nucleus).

In contrast, population 2 cells are larger and typical of primitive erythroid cells. **FRR2** fluorescence in circulating embryonic blood cells showed a pronounced decrease at 14.5 dpc, and by 15.5 dpc, fluorescence was almost at the levels seen in circulating adult erythrocytes that lack both nuclei and mitochondria.

Having observed differences between circulating embryonic and adult blood cells, it was important to investigate whether this was due to distinct erythropoietic (red blood cell production) environments or differences arising in the embryonic and adult blood streams. To address this, **FRR2** fluorescence was monitored in developing erythroid cells from the foetal liver just as it starts to operate as a haematopoietic organ (12.5 dpc), at the time when red blood cell production peaks (14.5 dpc). These results were compared to those for the adult bone marrow and the adult spleen. To specifically identify erythroid cells in these complex haematopoietic environments, **FRR2** was combined with antibody staining for multiparameteric flow cytometric analyses.

As discussed in Sect. 7.2.1, Ter-119 is an antibody recognising an antigen specific for developing and mature mouse erythroid cells. Samples from foetal liver, bone marrow and spleen were stained with a cocktail of **FRR2** and Ter-119 conjugated to Alexa Fluor™ 647. Ter-119-expressing cells were gated (Fig. 7.14) and assessed for **FRR2** fluorescence. No fluorescent signals were detected in unstained samples from the foetal liver, bone marrow or spleen (Fig. 7.14). In the 12.5 dpc foetal liver,

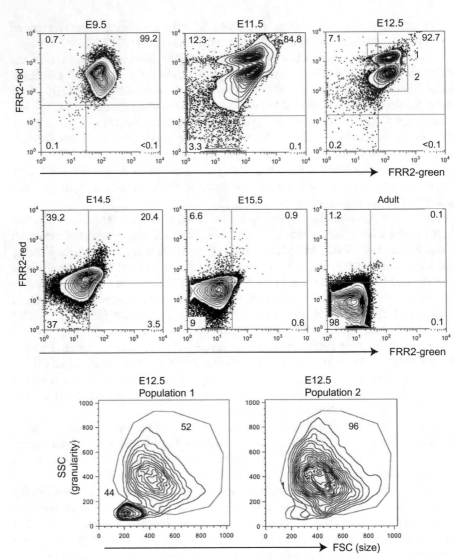

Fig. 7.13 Circulating blood cells show high levels of **FRR2**-green and **FRR2**-red fluorescence. As the foetus matures (14.5 dpc onwards), **FRR2** fluorescence is lost. By 15.5 dpc, most circulating blood cells resemble the adult circulating cells with essentially no **FRR2** fluorescence. At 12.5 dpc, two distinct populations can be detected according to **FRR2** fluorescence. Population 1 shows higher **FRR2**-red fluorescence compared to population 2. The bottom panel shows the size (forward scatter:FSC) versus cytoplasmic granularity (side scatter:SSC) of Populations 1 and 2 at 12.5 dpc. The numbers in the top right corner indicates the frequency of large, granular cells in all live circulating blood cells. The number in the bottom left corner represents the frequency of small, agranular cells in the live circulating blood cells. Flow cytometric profiles shown are representative of blood cells from 3–6 individual embryos at each stage and 6 adult mice. Reprinted with permission from *Antioxidants and Redox Signalling*, Volume 24, Issue 13, published by Mary Ann Leibert, Inc., New Rochelle, NY

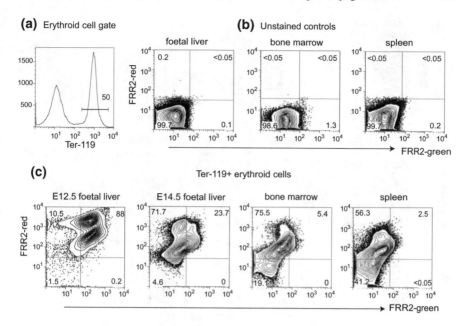

Fig. 7.14 Embryonic and adult blood production is distinguished by mitochondrial ROS levels. **a** Gating strategy for selection of Ter-119+ developing erythroid cells. The gate chosen is shown as the horizontal line. In (**a**), 50% of bone marrow cells are expressing Ter-119. **b** typical flow cytometry profiles of unstained foetal liver, bone marrow and spleen samples, **c** fluorescence of **FRR2**-green (488 nm) and **FRR2**-red (546 nm) fluorescence in early and later foetal liver erythroid cells compared to adult bone marrow and spleen Ter-119+ erythroid populations. Numbers in each corner refer to the frequency of live Ter-119+ cells in that quadrant. Flow cytometric profiles shown are representative of foetal livers from 3–6 individual embryos at each stage and bone marrow and spleens from 6 adult mice. Reprinted with permission from *Antioxidants and Redox Signalling*, Volume 24, Issue 13, published by Mary Ann Leibert, Inc., New Rochelle, NY

essentially all erythroid cells showed **FRR2** fluorescence (Fig. 7.14) and can be identified as two distinct populations, similar to the circulating blood at this stage (Fig. 7.13). By 14.5 dpc, **FRR2** fluorescence had dropped significantly in the foetal liver. In particular, the **FRR2**-bright population was significantly reduced. Within the adult bone marrow, **FRR2** fluorescence was further reduced, or restricted to the **FRR2**-red signal. **FRR2**-negative erythroid cells comprised close to half of all erythroid cells in the adult spleen. These data suggest that red blood cell production in the early embryo is profoundly different from the steady-state erythropoiesis found in the adult tissues.

To complete the analysis of mitochondrial ROS in hematopoietic cell types, **FRR2** fluorescence was assessed in developing and mature T-lymphocytes, macrophages, megakaryocytes and mast cells. Firstly, **FRR2** fluorescence was interrogated in all haematopoietic cells of the adult mouse bone marrow (Fig. 7.15a). Three populations could be observed (labelled 1, 2 and 3 in panel (a)). When gated for further analysis, each population showed distinct size and granularity profiles. Population 1, which

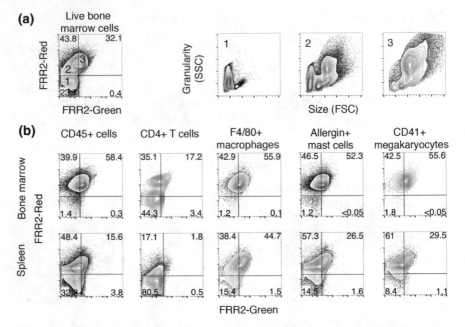

Fig. 7.15 CD4+ Helper T cells exhibit reduced mitochondrial ROS. **a** *Left panel* **FRR2**-red and green fluorescence on total live, gated bone marrow cells identifies three distinct populations (labelled 1, 2 and 3). Right panels show size (forward scatter, FSC) and granularity (side scatter, SSC) for populations 1, 2 and 3 as gated in left panel. **b** Live bone marrow and spleen cells were stained with antibodies against specific surface markers. Cells positive for the surface marker were then assess for **FRR2** fluorescence. The number shown in the corners represent the frequency of live gated cells within that quadrant. Data shown is representative of 3 independent experiments. Reprinted with permission from *Antioxidants and Redox Signalling*, Volume 24, Issue 13, published by Mary Ann Leibert, Inc., New Rochelle, NY

showed no **FRR2** fluorescence signal, consisted of very small, non-granular cells and are therefore most likely erythrocytes. Population 2, which showed low levels of **FRR2**-red fluorescence and no **FRR2**-green signal, was composed of a range of cells including some showing the size and granularity profiles of erythrocytes, erythroblasts (developing red blood cells) and myeloblasts (early myeloid cells). Population 3, which showed both **FRR2**-red and **FRR2**-green signal simultaneously, consisted largely of granular myeloid cells.

To better define the mitochondrial ROS in specific lineages, **FRR2** fluorescence was combined with antibodies against specific haematopoietic lineages. Adult mouse bone marrow and spleen were analysed to determine changes between haematopoietic cells where they develop (bone marrow) and where they are mature and functional (spleen). All haematopoietic cells, except for enucleated red blood cells, express the haematopoietic-specific surface marker CD45. CD45-expressing cells were selected by gating (Fig. 7.15b) and analysed for **FRR2** fluorescence. Bone marrow CD45+ showed uniform levels of **FRR2**-red and green fluorescence, but a significant population of CD45+ cells in the spleen had lost all **FRR2** fluorescence. Next, investigations focussed on determining which CD45+ cell type lost **FRR2** fluorescence once mature. CD4 is expressed by T-helper lymphocytes. Bone marrow CD4+ cells showed a population with mostly **FRR2**-red fluorescence and a second population lacking **FRR2** fluorescence. This second population (**FRR2**-negative) made up most of the CD4+ cells in the spleen, In contrast, F4/80+ macrophages, Allergin+ mast cells and CD41+ megakaryocytes showed uniform **FRR2**-fluorescence in the bone marrow and only moderate reduction in **FRR2** in the spleen. These data suggest that CD4+ T-helper lymphocytes in the spleen exhibit very low levels of mitochondrial ROS.

7.3.1 Conclusions

These experiments demonstrate that **FRR2** is capable of reporting on the mitochondrial oxidative capacities of haematopoietic cells during different stages of embryonic development and in the adult. Furthermore, **FRR2** could also identify the variations in the mitochondrial redox state of erythroid cells derived from haematopoietic organs such as the embryonic liver, and adult bone marrow and spleen. Amongst the haematopoietic cells that develop in the adult bone marrow, each cell type exhibited a distinct **FRR2**-fluorescence profile suggesting distinct mitochondrial ROS levels. Further investigations aimed towards understanding the correlation between the functions of these bone marrow-derived haematopoietic cells and their mitochondrial ROS levels would be valuable. Ex vivo studies such as these are particularly well suited to the use of small molecule probes like **NpFR2** and **FRR2** over genetically-encoded sensors.

References

1. G. Kaur, J.M. Dufour, Cell lines. Spermatogenesis **2**, 1–5 (2012)
2. J.R. Masters, G.N. Stacey, Changing medium and passaging cell lines. Nat. Protocols **2**, 2276–2284 (2007)
3. M. Lacroix, Persistent use of false cell lines. Int. J. Cancer **122**, 1–4 (2008)
4. J.-P. Gillet, S. Varma, M.M. Gottesman, The clinical relevance of cancer cell lines. J. Natl. Cancer Inst. **105**, 452–458 (2013)
5. I. Garitaonandia, H. Amir, F.S. Boscolo, G.K. Wambua, H.L. Schultheisz, K. Sabatini, R. Morey, S. Waltz, Y.-C. Wang, H. Tran, T.R. Leonardo, K. Nazor, I. Slavin, C. Lynch, Y. Li, R. Coleman, I. Gallego Romero, G. Altun, D. Reynolds, S. Dalton, M. Parast, J.F. Loring, L.C. Laurent, Increased risk of genetic and epigenetic instability in human embryonic stem cells associated with specific culture conditions. PLoS ONE **10**, e0118307 (2015)
6. C. Pan, C. Kumar, S. Bohl, U. Klingmueller, M. Mann, Comparative proteomic phenotyping of cell lines and primary cells to assess preservation of cell type-specific functions. Mol. Cellular Proteomics **8**, 443–450 (2009)

7. S. Wilkening, F. Stahl, A. Bader, comparison of primary human hepatocytes and hepatoma cell line HEPG2 with regard to their biotransformation properties. Drug Metab. Dispos. **31**, 1035–1042 (2003)
8. V. Sanchez-Valle, N.C. Chavez-Tapia, M. Uribe, N. Mendez-Sanchez, Role of oxidative stress and molecular changes in liver fibrosis: a review. Curr. Med. Chem. **19**, 4850–4860 (2012)
9. C. Guguen-Guillouzo, A. Guillouzo, General review on in vitro hepatocyte models and their applications, in *Methods in Molecular Biology*, ed. by N.J. Clifton, vol. 640 (2010), pp. 1–40
10. C. Garcia-Ruiz, J.C. Fernandez-Checa, Redox regulation of hepatocyte apoptosis. J. Gastroenterol. Hepatol. **22**(Suppl 1), S38–42 (2007)
11. R. Singh, M.J. Czaja, Regulation of hepatocyte apoptosis by oxidative stress. J. Gastroenterol. Hepatol. **22**(Suppl 1), S45–8 (2007)
12. I. Kurose, H. Higuchi, S. Miura, H. Saito, N. Watanabe, R. Hokari, M. Hirokawa, M. Takaishi, S. Zeki, T. Nakamura, H. Ebinuma, S. Kato, H. Ishii, Oxidative stress-mediated apoptosis of hepatocytes exposed to acute ethanol intoxication. Hepatology **25**, 368–378 (1997)
13. Y. Sumida, E. Niki, Y. Naito, T. Yoshikawa, Involvement of free radicals and oxidative stress in NAFLD/NASH. Free Radic. Res. **47**, 869–880 (2013)
14. M.D. Norenberg, A.R. Jayakumar, K.V. Rama, Rao, oxidative stress in the pathogenesis of hepatic encephalopathy. Metab. Brain Dis. **19**, 313–329 (2004)
15. H. Tsukamoto, Oxidative stress, antioxidants, and alcoholic liver fibrogenesis, in *Alcohol (Fayetteville, N.Y.)*, vol. 10 (1993), pp. 465–467
16. S. Pal, S.J. Polyak, N. Bano, W.C. Qiu, R.L. Carithers, M. Shuhart, D.R. Gretch, A. Das, Hepatitis C virus induces oxidative stress, DNA damage and modulates the DNA repair enzyme NEIL1. J. Gastroenterol. Hepatol. **25**, 627–634 (2010)
17. H. Cichoż-Lach, A. Michalak, Oxidative stress as a crucial factor in liver diseases. World J. Gastroenterol. WJG **20**, 8082–8091 (2014)
18. B. Saberi, M. Shinohara, M.D. Ybanez, N. Hanawa, W.A. Gaarde, N. Kaplowitz, D. Han, Regulation of H(2)O(2)-induced necrosis by PKC and AMP-activated kinase signaling in primary cultured hepatocytes. Am. J. Physiol. Cell Physiol. **295**, C50–63 (2008)
19. M.G. Cotticelli, A.M. Crabbe, R.B. Wilson, M.S. Shchepinov, Insights into the role of oxidative stress in the pathology of Friedreich ataxia using peroxidation resistant polyunsaturated fatty acids. Redox Biol. **1**, 398–404 (2013)
20. S. Hill, C.R. Lamberson, L. Xu, R. To, H.S. Tsui, V.V. Shmanai, A.V. Bekish, A.M. Awad, B.N. Marbois, C.R. Cantor, N.A. Porter, C.F. Clarke, M.S. Shchepinov, Small amounts of isotope-reinforced polyunsaturated fatty acids suppress lipid autoxidation. Free Radic. Biol. Med. **53**, 893–906 (2012)
21. L.A. Herzenberg, D. Parks, B. Sahaf, O. Perez, M. Roederer, L.A. Herzenberg, The history and future of the fluorescence activated cell sorter and flow cytometry: a view from stanford. Clin. Chem. **48**, 1819–1827 (2002)
22. Regenerative Medicine, Technical report, Department of Health and Human Services
23. D. Levitt, R. Mertelsmann, *Hematopoietic Stem Cells: Biology and Therapeutic Applications* (Taylor & Francis, 1995)
24. C.J. Eaves, Hematopoietic stem cells: concepts, definitions, and the new reality. Blood **125**, 2605–2613 (2015)
25. I. Godin, A. Cumano, *Hematopoietic Stem Cell Development* (Medical Intelligence Unit, Springer, US, 2010)
26. M. Kondo, *Hematopoietic Stem Cell Biology*. Stem Cell Biology and Regenerative Medicine (Humana Press, 2009)
27. D. Hernandez-Garcia, C.D. Wood, S. Castro-Obregon, L. Covarrubias, Reactive oxygen species: a radical role in development? Free Radic. Biol. Med. **49**, 130–143 (2010)
28. H. Sandoval, P. Thiagarajan, S.K. Dasgupta, A. Schumacher, J.T. Prchal, M. Chen, J. Wang, Essential role for Nix in autophagic maturation of erythroid cells. Nature **454**, 232–235 (2008)
29. N.A. Maianski, J. Geissler, S.M. Srinivasula, E.S. Alnemri, D. Roos, T.W. Kuijpers, Functional characterization of mitochondria in neutrophils: a role restricted to apoptosis. Cell Death Differ. **11**, 143–153 (2004)

30. C. Nombela-Arrieta, G. Pivarnik, B. Winkel, K.J. Canty, B. Harley, J.E. Mahoney, S.-Y. Park, J. Lu, A. Protopopov, L.E. Silberstein, Quantitative imaging of haematopoietic stem and progenitor cell localization and hypoxic status in the bone marrow microenvironment. Nat. Cell Biol. **15**, 533–543 (2013)

31. H.M. Shapiro, *Practical Flow Cytometry* (Wiley, 2005)

32. S.T. Fraser, R.G. Midwinter, B.S. Berger, R. Stocker, Heme oxygenase-1: a critical link between iron metabolism, erythropoiesis, and development. Adv. Hematol. **2011**, 473709 (2011)

33. J. Isern, S.T. Fraser, Z. He, M.H. Baron, Developmental niches for embryonic erythroid cells. Blood Cells Molecules Dis. **44**, 207–208 (2010)

34. K. McGrath, J. Palis, Ontogeny of erythropoiesis in the mammalian embryo. Curr. Top. Dev. Biol. **82**, 1–22 (2008)

35. P.D. Kingsley, J. Malik, K.A. Fantauzzo, J. Palis, Yolk sac-derived primitive erythroblasts enucleate during mammalian embryogenesis. Blood **104**, 19–25 (2004)

36. S.T. Fraser, J. Isern, M.H. Baron, Maturation and enucleation of primitive erythroblasts during mouse embryogenesis is accompanied by changes in cell-surface antigen expression. Blood **109**, 343–352 (2006)

37. M. Socolovsky, Molecular insights into stress erythropoiesis. Curr. Opin. Hematol. **14**, 215–224 (2007)

38. M.H. Baron, Embryonic origins of mammalian hematopoiesis. Exp. Hematol. **31**, 1160–1169 (2003)

39. K. Ito, A. Hirao, F. Arai, S. Matsuoka, K. Takubo, I. Hamaguchi, K. Nomiyama, K. Hosokawa, K. Sakurada, N. Nakagata, Y. Ikeda, T.W. Mak, T. Suda, Regulation of oxidative stress by ATM is required for self-renewal of haematopoietic stem cells. Nature **431**, 997–1002 (2004)

40. Y.-Y. Jang, S.J. Sharkis, A low level of reactive oxygen species selects for primitive hematopoietic stem cells that may reside in the low-oxygenic niche. Blood **110**, 3056–3063 (2007)

41. P. Rimmelé, C. Bigarella, R. Liang, B. Izac, R. Dieguez-Gonzalez, G. Barbet, M. Donovan, C. Brugnara, J. Blander, D. Sinclair, S. Ghaffari, Aging-like phenotype and defective lineage specification in SIRT1-deleted hematopoietic stem and progenitor cells. Stem Cell Rep. **3**, 44–59 (2016)

Chapter 8
Non-mammalian Systems

While most biological investigations are performed using mammalian cells and tissues, non-mammalian systems are also commonly studied. Studying such systems essentially involves the use of simple animals models with striking similarities to mammalian systems, such as zebrafish (*Danio Rerio*) for developmental biology [1], *Drosophila* for genetics [2] and *C. Elegans* for nervous system [3], or investigating facets of health and disease indirectly related to mammals, such as antibiotic resistance resulting from the formation of bacterial biofilms [4]. Whilst there exist a wide array of fluorescent tools for use in mammalian systems [5, 6], not much attention has been paid towards development of new probes or the use of existing probes in non-mammalian systems. The aim of this section of the work was to perform investigations in non-mammalian systems using the fluorescent redox probes developed over the course of this project (Fig. 8.1), highlighting their ability to report on oxidative changes diverse biological systems.

8.1 Caenorhabditis Elegans

Nematodes, or roundworms, are a diverse group of free-living (non-parasitic) and parasitic worms that constitute the phylum Nematoda. Parasitic nematodes have been primarily studied to develop measures to prevent and cure the infections that they can cause. Recently, extensive biological investigations have been performed with the free-living *Caenorhabditis elegans* (*C. Elegans*). It is the simplest organism with a nervous system, and has been used as a model organism for studying neurology, developmental biology and aging following the first study by Dr Sydney Brenner in 1963 [7].

Although the organism is small (about 1 mm in length) and primitive, *C. Elegans* shares several fundamental biological characteristics and cellular structures with more highly developed organisms. With a short life cycle (3 days) and a diet of bac-

© Springer International Publishing AG 2018
A. Kaur, *Fluorescent Tools for Imaging Oxidative Stress in Biology*,
Springer Theses, https://doi.org/10.1007/978-3-319-73405-7_8

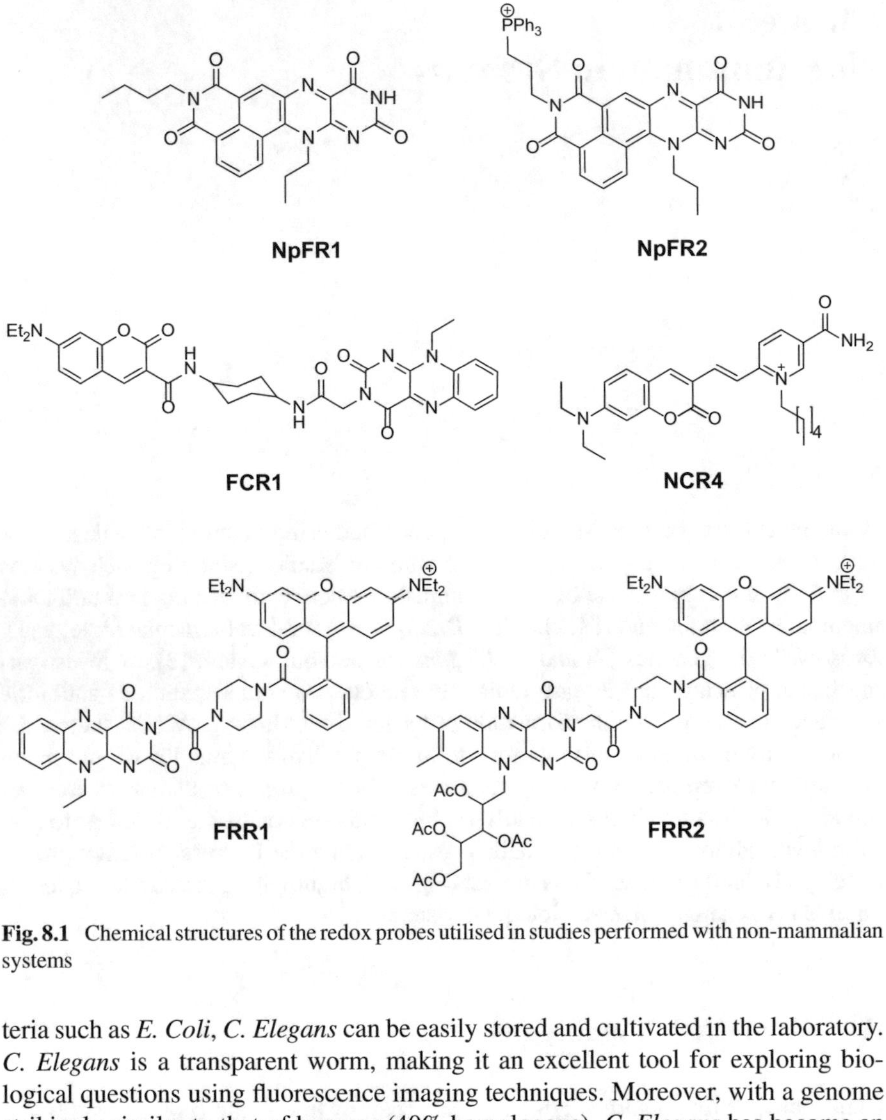

Fig. 8.1 Chemical structures of the redox probes utilised in studies performed with non-mammalian systems

teria such as *E. Coli*, *C. Elegans* can be easily stored and cultivated in the laboratory. *C. Elegans* is a transparent worm, making it an excellent tool for exploring biological questions using fluorescence imaging techniques. Moreover, with a genome strikingly similar to that of humans (40% homologous), *C. Elegans* has become an interesting organism to investigate human diseases. Owing to these characteristics, *C. Elegans* has been a powerful research organism for almost 5 decades [8, 9]

8.1.1 Probe Uptake Studies

The aim of this work was to develop fluorescent redox probes that would applicable to a wide variety of biological systems. All the redox probes developed thus far;

both cytoplasmic (**NpFR1** and **FCR1**) and mitochondrial (**NpFR2, FRR1, FRR2** and **NCR4**), were investigated for their uptake in *C. Elegans*. These studies were performed in the laboratory of Dr Gawain McColl, Head of Molecular Gerontology Laboratory, The Florey Institute of Neuroscience and Mental Health.

The probe uptake studies were performed using the wild-type Bristol strain (N2) of *C. Elegans*, which were maintained on lawns of *Escherischia coli OP 50* strain, a slow growing uracil-auxotroph. The adult worms were filtered using a 40 μm nylon mesh and incubated in different tubes containing 100 μM of each probe in S basal medium (a liquid medium to maintain *C. Elegans*) containing *E. Coli* OP 50 as a food source. The worms were incubated at 25 °C for 1 h with constant mixing. It was anticipated that while feeding on the bacteria in the medium, the worms would also ingest the probe. Following incubation, the worms were washed and allowed to rest in probe-free S basal medium for 20 min to flush out any probe remaining in the digestive tract of the worms. The worms were then transferred onto glass slides coated with 3% agar pads containing 2% sodium azide. The sodium azide immobilises the worms, aiding their imaging.

Confocal microscopy studies confirmed the uptake of all the probes in *C. Elegans*. It was observed that the cytoplasmic probes (**NpFR1** and **FCR1**) were able to diffuse into deeper tissues away from the digestive tract of the worms, whereas the mitochondrial probes (**FRR1, FRR2** and **NCR4**) were highly localised around the gut of the worms (Fig. 8.2).

This could be a consequence of paralysing the worms using sodium azide, a mitochondrial uncoupler [10]. As a result in future studies other alternatives such as levamisole should be used [11].

8.1.2 Oxidative Stress and Heat Shock in C. Elegans

Signalling pathways involved in resistance towards stress are highly conserved evolutionarily, particularly those responsible for dealing with oxidative stress [12, 13]. Although many experiments have been reported in Chaps. 6 and 7 that measure oxidative stress and the corresponding protective mechanisms in a variety of cellular models, *in cellulo* studies only partially reproduce organismal conditions. Therefore, the study of oxidative stress in model organisms such as *C. Elegans* has great significance, particularly because the results obtained could be translated into better understanding of human pathological conditions associated with oxidative stress [14, 15].

C. Elegans are typically cultured at 20–25 °C. High-temperature conditions, of 30–35 °C constitute thermal stress, or heat shock, which results in the upregulation of heat-shock proteins [16]. Acute thermal stress conditions in worms have been reported to reduce fertility and accelerate the process of aging [16, 17]. Morever, it has been observed that treatment with antioxidant compounds, such as α-lipoic acid and trolox, increases thermal tolerance and prevents early aging in *C. Elegans*, suggesting that oxidative stress is one of the downstream effects of thermal shock

Fig. 8.2 Confocal microscopy images of *C. Elegans* treated with flavin-based redox probes **a** vehicle control **b NpFR1** (100 μM, 1 h, $\lambda_{ex} = 405$ nm, $\lambda_{em} = 480 - 600$ nm), **c, d FCR2** (100 μM, 1 h, $\lambda_{ex} = 820$ nm) in blue ($\lambda_{em} = 420 - 475$ nm) and green ($\lambda_{em} = 520 - 600$ nm) channels, **e FRR1** (100 μM, 1 h, $\lambda_{ex} = 488$ nm), $\lambda_{em} = 560 - 650$ nm), **f FRR2** (100 μM, 1 h, $\lambda_{ex} = 488$ nm), $\lambda_{em} = 560 - 650$ nm) and **NCR4** (100 μM, 1 h, $\lambda_{ex} = 458$ nm) in (**g**) channel 1 ($\lambda_{em} = 470 - 580$ nm) and **h** channel 2 ($\lambda_{em} = 600 - 750$ nm). Scale bar represents 50 μm

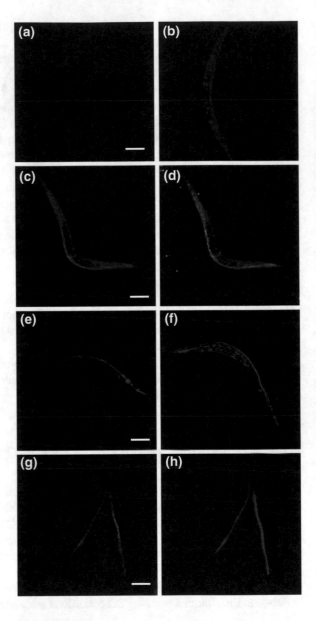

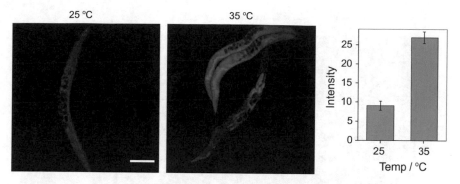

Fig. 8.3 Confocal microscopy images of *C. Elegans* treated **NpFR1** (λ_{ex} = 405 nm, λ_{em} = 480 – 600 nm) and incubated at 25 and 35 °C. The bar graph on the right represents average intensity per worm when incubated at given temperatures. Scale bar represents 100 μm

[16, 18]. In contrast, antioxidants such as *N*-acetylcysteine and vitamin C are not effective in protecting *C. Elegans* against the effects of thermal shock [19].

Having confirmed the uptake of the probes in *C. Elegans*, the ability of the probes to respond to changes in oxidative capacity or copper levels in thermally stressed worms was investigated. *C. Elegans* were incubated in S basal medium (containing *E. Coli* OP 50 and 100 μM of the probes) at 35 °C for a duration of 1 h. A control set of worms were maintained at 25 °C.

Following incubation, the worms were washed and prepared for imaging. Both the cytoplasmic redox probes **NpFR1** (Fig. 8.3) and **FCR1** (Fig. 8.4) report an increased oxidative capacity in heat shocked worms. **NpFR1** fluorescence was 3-fold higher in worms incubated at 35 °C compared to those at 25 °C, and **FCR1**-treated worms exhibited a 10-fold increase in the ratio of green/blue emission in the case of thermally stressed worms.

While mitochondrial probes exhibited poor penetration into deeper tissues, they nevertheless exhibited similar response to heat shock and their redox responsive abilities remained intact. In heat-shocked *C. Elegans* **NpFR2** reported approximately 20-fold higher fluorescence intensity, whilst those treated with **FRR1** and **FRR2** showed 2-fold higher rhodamine emission (Fig. 8.5).

The nicotinamide-based redox probe **NCR4**, failed to show any significant changes in fluorescence response (green/red emission ratio, Fig. 8.6), which could be due to the irreversible reduction of this probe.

8.1.3 Conclusions

These experiments establish *C. Elegans* as a simple model to investigate chemical changes within an organism. This transparent worm allowed for whole animal imaging without complications related to light penetration or interference from

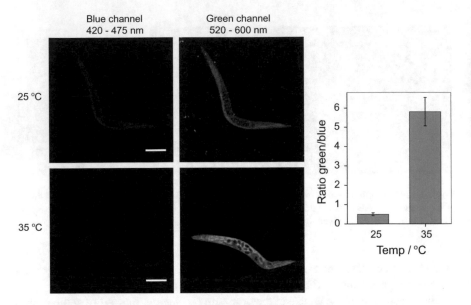

Fig. 8.4 Confocal microscopy images of *C. Elegans* treated with **FCR1** ($\lambda_{ex} = 405$ nm) collected in both blue and green channels and incubated at 25 and 35 °C. The bar graph on the right represents ratio of green/blue intensity per worm when incubated at given temperatures. Scale bar represents 100 μm

physiological chromophores and light absorbing lipids. The investigations expand the scope of small molecule fluorescent probes, demonstrating their ability to image non-mammalian biological systems. The developed redox probes discussed in Chaps. 2–5, as well as probes developed by other members of the research group, show excellent potential to report on chemical variations, whether oxidative capacity or copper pools in *C. Elegans*. Having established the accumulation and successful visualisation of these fluorescent probes as well their sensing abilities, the probes can be employed in future investigations aiming towards answering complex biological questions with the use of appropriate mutants of *C. Elegans*.

8.2   Bacterial Biofilms

Acute infections caused by pathogenic bacteria have been studied extensively for centuries. Modern control measures such as vaccines and antibiotics have been very effective in combating such infections. Bacterial growth can be classified into two modes–planktonic and biofilms [20]. It has only been in the past 15 years that the medical community has recognised that biofilm bacteria as fundamentally distinct from planktonic bacteria [21]. Planktonic bacteria (existing as single independent cells) are responsible for a variety of acute infections that can generally be cured

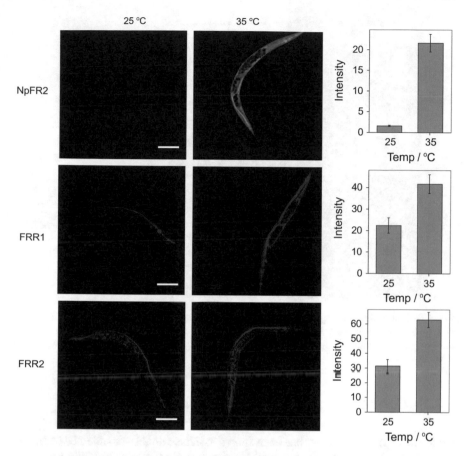

Fig. 8.5 Confocal microscopy images of *C. Elegans* treated with mitochondrial probes **NpFR2** ($\lambda_{ex} = 405$ nm, $\lambda_{em} = 480 - 600$ nm) and **FRR1** and **FRR2** ($\lambda_{ex} = 488$ nm, $\lambda_{em} = 560 - 650$ nm)incubated at 25 and 35 °C. The bar graph on the right represents average intensity per worm when incubated at given temperatures. Scale bar represents 100 μm

with antibiotics, but chronic infections are induced by bacteria growing in slime-enclosed aggregates known as biofilms, which act as a multicellular organism. Bacterial biofilms can defend themselves against human immune cells or antibiotics, making them difficult to eradicate, and resulting in several chronic infections [22–24].

Within a biofilm, bacteria grow in highly crowded and competitive environments with limited supply of nutrients. These stresses elicit a myriad of highly regulated adaptive processes that not only shield bacteria from the exogenous stress, but also manifest biochemical changes that affect antimicrobial vulnerability [25]. Exposure to limited nutrient supply, ROS/RNS (oxidative/nitrosative stress), elevated temperature (thermal stress) and damaged membrane (envelope stress) promote physiological and biochemical changes that compromise antimicrobial activity [26, 27].

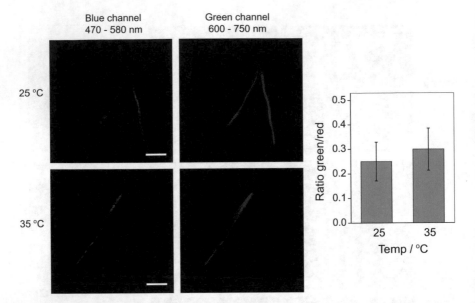

Fig. 8.6 Confocal microscopy images of *C. Elegans* treated **NCR4** (λ_{ex} = 458 nm) collected in both green and red channels and incubated at 25 and 35 °C. The bar graph on the right represents ratio of green/red intensity per worm when incubated at given temperatures. Scale bar represents 100 μm

8.2.1 Oxidative Stress in Bacterial Biofilms

Endogenous oxidative stress has been shown to induce double-stranded DNA breaks that result in mutations, which contribute significantly towards antibiotic resistance of the biofilms [28]. Furthermore, it is possible that the deeper residents of the biofilm experience greater stress levels compared to the ones closer to the surface. It was envisioned that the fluorescent redox probes developed during the course of this PhD would be able to provide valuable information about the variations in oxidative stress levels with the depth of the biofilm. To be able to perform this investigation, it is crucial that the probes are capable of penetrating the biofilm and accumulating within the bacterial cells.

NpFR1 was investigated for uptake in bacterial biofilms, and these studies were performed in collaboration with Dr Anahit Penesyan, Department of Chemistry and Biomolecular Sciences, Macquarie University. Biofilms were prepared on glass slides by Dr Penesyan from Gram-positive (*Bacillus cereus*-BC) and Gram-negative (*Acinetobacter baumannii*-AB and *Pseudomonas aeruginosa*-PA) bacterial strains using continuous flow-cell system. The biofilms were then maintained in LB medium containing **NpFR1** (50 μM) for 15 min followed by washing with probe-free LB medium for 10 min after which the slides were imaged using a confocal microscope. A control slide was prepared by treatment with Live/Dead® BacLight™ stain, a mixture of

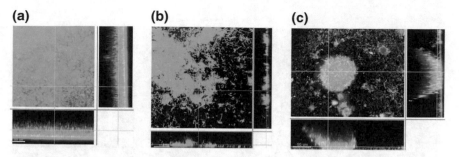

Fig. 8.7 Confocal microscopy z-stacks of bacterial biofilms cultured from strains **a** *Acinetobacter baumannii* **b** *Bacillus cereus* and **c** *Pseudomonas aeruginosa* treated with Live/Dead® BacLight™ stain (λ_{ex} = 488 nm). Green pixels represent live cells and yellow pixels represent dead cells

SYTO® 9 nucleic acid stain, which permeates live cells and fluoresces green upon binding to the DNA, and propidium iodide a red-fluorescent dye, which cannot permeate live cells but enters dead cells. As a result, live cells fluoresce green but dead cells fluoresce green and red, and therefore appear yellow.

As shown in Fig. 8.7, all three strains exhibited a good uptake of the Live/Dead® BacLight™ stain. AB and BC strains exhibited lawn-like characteristics whilst PA formed densely packed microcolonies (Fig. 8.7). Such bacterial microcolonies could be particularly interesting because of the possible existence of a gradient in oxygen availability with increasing depth of the colony.

For AB and BC strains, the penetrating ability of **NpFR1** was observed to be a few microns deep (Fig. 8.8). In contrast, PA showed the presence of distinct microcolonies (densely packed spherical structures) formed by the bacteria (Fig. 8.9) and the images obtained using the Live/Dead® BacLight™ staining suggest a great proportion of dead cells in the centre of the microcolony compared to its periphery (Fig. 8.7c).

The biofilms formed from the PA strain showed bright **NpFR1** fluorescence suggesting a good level of probe uptake in the microcolonies (Fig. 8.9c). Owing to its ability of form microcolonies and the permeability of **NpFR1**, these studies establish *Pseudomonas aeruginosa* as a good system for investigating the applicability of redox probes.

The fluorescence intensity **NpFR1** was observed to decrease moving inwards to the centre of the microcolony (Fig. 8.9c). However, at this stage it was difficult to conclude whether this decrease in fluorescence intensity is a consequence of lower levels of probe uptake or the presence of reduced form of the probe. This can be overcome by employing a ratiometric redox probe. Similar investigations were performed on the biofilms produced from the PA strain on another day using a different experimental setup. The biofilms were incubated with LB medium supplemented with 20 μM **FCR1** and prepared for imaging. The biofilms treated with **FCR1** demonstrated very superficial probe uptake and the formation of microcolonies was not observed, with the biofilm instead exhibiting lawn-like characteristics. Images were recorded in both blue (420–470 nm) and green (520–600 nm) channels upon 405 nm excitation and were used to obtain ratio images (green/blue) using the RatioPlus plugin of Fiji.

(a)   **(b)**

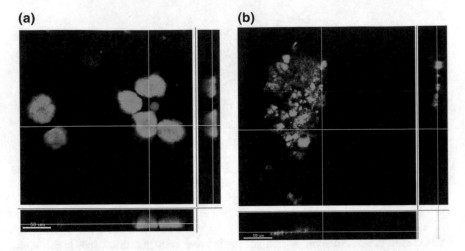

Fig. 8.8 Confocal microscopy z-stacks of biofilms cultured from bacterial strains *Acinetobacter baumannii* (left) and *Bacillus cereus* (right) treated with **NpFR1** (λ_{ex} = 488 nm)

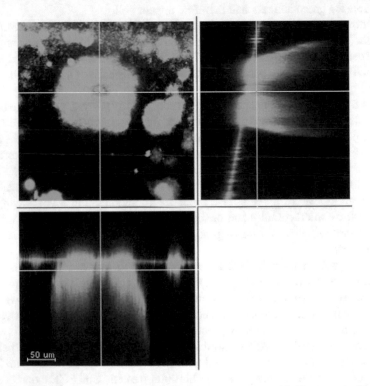

Fig. 8.9 Confocal microscopy z-stacks of *Pseudomonas aeruginosa* biofilm treated with **NpFR1** (λ_{ex} = 488 nm) showing dense microcolonies and a gradient of **NpFR1** fluorescence

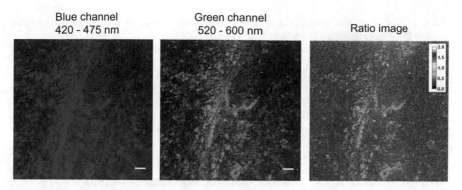

Fig. 8.10 Confocal microscopy images of *Pseudomonas aeruginosa* biofilm treated with **FCR1** (λ_{ex} = 405 nm) collected in both blue and green channels. The ratio image represent a ratio of the green/blue intensity. Scale bars represent 50 μM

The ratio image suggest a slightly higher ratio, corresponding to higher oxidative capacity in the cells present in the middle of the bacterial lawn (Fig. 8.10), suggesting that the bacterial cells residing at the core of the biofilm have higher oxidative stress levels than those at the periphery.

8.2.2 Conclusions

The results obtained from these experiments demonstrate that **NpFR1** and **FCR1** can be used for imaging redox within the unicellular bacteria. Greater extent of probe accumulation was observed in case of gram negative bacterial strains compared to the gram positive counterparts. Furthermore, **NpFR1** and **FCR1** reported that the biofilms developed from *Pseudomonas aeruginosa* exhibited a decrease in the oxidative capacity towards the centre of the biofilm. Future investigations could be performed to investigate the effect of various antibiotics on the formation and oxidative status of biofilms formed from *Pseudomonas aeruginosa*.

References

1. W. Driever, D. Stemple, A. Schier, L. Solnica-Krezel, Zebrafish: genetic tools for studying vertebrate development. Trends Genet. **10**, 152–159 (1994)
2. J.R. Powell, *Progress and Prospects in Evolutionary Biology: The Drosophila Model: The Drosophila Model* (Oxford University Press, USA, 1997)
3. O. Bloom, Non-mammalian model systems for studying neuro-immune interactions after spinal cord injury. Exp. Neurol. **258**, 130–140 (2014)
4. G.G. Anderson, G.A. O'Toole, *Bacterial Biofilms, Chapter Innate and (85–105)* (Springer, Heidelberg, 2008)

5. X. Li, X. Gao, W. Shi, H. Ma, Design strategies for water-soluble small molecular chromogenic and fluorogenic probes. Chem. Rev. **114**, 590–659 (2013)
6. Z. Guo, S. Park, J. Yoon, I. Shin, Recent progress in the development of near-infrared fluorescent probes for bioimaging applications. Chem. Soc. Rev. **43**, 16–29 (2014)
7. S. Brenner, The genetics of Caenorhabditis elegans. Genetics **77**, 71–94 (1974)
8. E.K. Marsh, R.C. May, Caenorhabditis elegans, a model organism for investigating immunity. Appl. Environ. Microbiol. **78**, 2075–2081 (2012)
9. M.C.K. Leung, P.L. Williams, A. Benedetto, C. Au, K.J. Helmcke, M. Aschner, J.N. Meyer, Caenorhabditis elegans: an emerging model in biomedical and environmental toxicology. Toxicol. Sci. **106**, 5–28 (2008)
10. L. Zhang, L. Li, L. Ban, W. An, S. Liu, X. Li, B. Xue, Y. Xu, Effect of sodium azide on mitochondrial membrane potential in SH-SY5Y human neuroblastoma cells. Zhongguo yi xue ke xue yuan xue bao. Acta Academiae Medicinae Sinicae **22**, 436–439 (2000)
11. R.J. Martin, A.P. Robertson, S.K. Buxton, R.N. Beech, C.L. Charvet, C. Neveu, Levamisole receptors: a second awakening. Trends Parasitol. **28**, 289–296 (2012)
12. J.M. Van Raamsdonk, S. Hekimi, Reactive oxygen species and aging in caenorhabditis elegans: causal or casual relationship? Antioxid. Redox Signal. **13**, 1911–1953 (2010)
13. R. Baumeister, E. Schaffitzel, M. Hertweck, Endocrine signaling in Caenorhabditis elegans controls stress response and longevity. J. Endocrinol. **190**, 191–202 (2006)
14. M. Markaki, N. Tavernarakis, Modeling human diseases in Caenorhabditis elegans. Biotechnol. J. **5**, 1261–1276 (2010)
15. M. Rodriguez, L.B. Snoek, M. De Bono, J.E. Kammenga, Worms under stress: C. elegans stress response and its relevance to complex human disease and aging. Trends in genetics. TIG **29**, 367–374 (2013)
16. K.I. Zhou, Z. Pincus, F.J. Slack, Longevity and stress in Caenorhabditis elegans. Aging (Albany NY) **3**, 733–753 (2011)
17. C. Portal-Celhay, E.R. Bradley, M.J. Blaser, Control of intestinal bacterial proliferation in regulation of lifespan in Caenorhabditis elegans. BMC Microbiol. **12**, 49 (2012)
18. R.P. Oliveira, J. Porter Abate, K. Dilks, J. Landis, J. Ashraf, C.T. Murphy, T.K. Blackwell, Condition-adapted stress and longevity gene regulation by Caenorhabditis elegans SKN-1/Nrf. Aging Cell **8**, 524–541 (2009)
19. D. Gems, R. Doonan, Antioxidant defense and aging in C. elegans: is the oxidative damage theory of aging wrong? Cell cycle (Georgetown, Tex.) **8**, 1681–1687 (2009)
20. M. Kostakioti, M. Hadjifrangiskou, S.J. Hultgren, Bacterial biofilms: development, dispersal, and therapeutic strategies in the dawn of the postantibiotic era, in *Cold Spring Harbor Perspectives in Medicine*, vol. 3 (2013)
21. P. Watnick, R. Kolter, Biofilm, city of microbes. J. Bacteriol. **182**, 2675–2679 (2000)
22. R.M. Donlan, Biofilms: microbial life on surfaces. Emerg. Infect. Dis. J. **8**, 881 (2002)
23. T. Bjarnsholt, The role of bacterial biofilms in chronic infections. APMIS. Supplementum, 1–51 (2013)
24. R.M. Donlan, Biofilm formation: a clinically relevant microbiological process. Clin. Infect. Dis. **33**, 1387–1392 (2001)
25. B.R. Boles, P.K. Singh, Endogenous oxidative stress produces diversity and adaptability in biofilm communities. Proc. Natl. Acad. Sci. U.S.A. **105**, 12503–12508 (2008)
26. N. Høiby, T. Bjarnsholt, M. Givskov, S. Molin, O. Ciofu, Antibiotic resistance of bacterial biofilms. Int. J. Antimicrob. Agents **35**, 322–332 (2010)
27. C.A. Fux, J.W. Costerton, P.S. Stewart, P. Stoodley, Survival strategies of infectious biofilms. Trends Microbiol. **13**, 34–40 (2005)
28. K. Poole, Bacterial stress responses as determinants of antimicrobial resistance. J. Antimicrob. Chemother. **67**(9), 2069–2089 (2012)

Chapter 9
Conclusions

The design and development of small molecules fluorescent probes for redox sensing applications has received extensive interest and attention [1, 2]. Reviews of the literature in this domain highlight that much of the research attention has been focussed towards the development of irreversible reaction based probes selective for specific ROS/RNS, and not much attention has been paid towards the development of reversible redox probes that can report on the changes in redox environment with temporal resolution [3–6]. Furthermore, probe design strategies thus far paid little attention to tuning the redox potential of the developed probes to suit the biological window.

Most approaches have been biased towards incorporating a fluorescence-quenching moiety on a fluorophore scaffold that is either modified or cleaved upon reaction with specific ROS/RNS. In contrast, investigation of the use of cell's own redox-active molecules for the purpose of fluorescence redox sensing has been underestimated, and warrants greater consideration [6]. Moreover, the attributes of the fluorescence change upon oxidation/reduction are pivotal in determining the biological utility of a probe. In addition to minimising interference from background effects, ratiometric probes offer the possibility of imaging both oxidised and reduced forms, and therefore possess great potential to enable quantification of variations in relative or absolute redox potential.

The investigations reported in this thesis explore two different naturally occurring redox-active molecules, flavins and nicotinamides for use as fluorescent redox sensors. Different strategies of developing flavin-based redox sensors with a ratiometric output have been discussed in Chaps. 2–4. The first strategy involved the incorporation of a different fluorophore, such as naphthalimide within the flavin scaffold, which would exhibit its own fluorescence properties upon the reduction of the flavin. This approach was investigated in the design of **NpFR1**, discussed in Chap. 2, but it was successful only at shorter excitation wavelengths. Nevertheless, at longer excitation wavelengths **NpFR1** demonstrated red-shifted fluorescence properties compared to cellular flavins, enabling distinct identification of its fluorescence response

© Springer International Publishing AG 2018

A. Kaur, *Fluorescent Tools for Imaging Oxidative Stress in Biology*,

Springer Theses, https://doi.org/10.1007/978-3-319-73405-7_9

from natural flavins. Despite such marked structural modifications to the flavin scaffold, **NpFR1** exhibited excellent reversible sensing properties to reduction-oxidation cycles and a biologically-relevant redox potential.

Another strategy explored for designing ratiometric probes was FRET. Chapter 3 details the approaches employed to develop an emission ratiometric probe for which flavin was fixed as a FRET-acceptor absorbing energy from an appropriate donor fluorophore, such as coumarin. Different synthetic schemes were investigated using the acetylated form of the naturally existing riboflavin and the synthetic *N*-ethylflavin. It was understood that the labile acetyl groups on the tetraacetylriboflavin posed difficulties towards the synthesis and purification, therefore for all future directions involving probe design it is suggested to use *N*-ethylflavin to investigate the proof of principle before attempting a challenging synthesis with the tetraacetylriboflavin.

Synthetic methodology involving the commonly employed copper catalysed click chemistry was found to be unsuccessful due to possible interferences from redox-active flavin moiety. It might therefore be valuable to use ruthenium-based catalysts for click chemistry reactions in the future as these are not perturbed by the presence of redox-active molecules such as flavins [7, 8]. Finally, a synthetic procedure involving the use of the non-aromatic rigid linker *trans*-1,4-diaminocyclohexane to tether flavin and coumarin through the formation of amide bonds on either ends of the linker proved successful. The resulting probe **FCR1** demonstrated excellent fluorescence properties with a FRET-based ratiometric response in its emission profile towards changes in the redox environment. Although the reversibility of the flavin scaffold in **FCR1** was compromised, the reasons for which remain unclear and warrant further investigation, nonetheless despite tethering a whole fluorophore scaffold to the flavin molecule, the redox potential of **FCR1** lay within the biological window and the sensor demonstrated excellent redox responsive behaviour both in vitro and in vivo.

The manifest applicability of the developed fluorescent redox probes in biological systems requires simultaneous fulfilment of a number of properties. The probe must exhibit excellent photophysical properties, particularly with the use of excitation sources that are not detrimental to the cells. An ideal probe candidate would be non-cytotoxic and cause minimal perturbation to the cellular redox chemistry, for which it is essential that the redox potential of the probe be well within the biological range. Furthermore, the sensor molecule should possess suitable sites on the scaffold that can be righteously modified to deliver the probe to sub-cellular organelles of interest without hampering its photophysical and redox-sensing properties. Each of the probes developed in this work was assessed according to these criteria.

Cytotoxicity studies and biological experiments performed to establish the redox-responsive abilities of probes in simple biological systems ensured that both **NpFR1** and **FCR1** satisfied the desired criteria and demonstrated that the probes mainly localised in the cytoplasm. Moreover, these studies highlighted the applicability of **FCR1** in different imaging modalities enabling it use in a variety of biological domains. The cytoplasmic **NpFR1** and **FCR1** were suitably modified to direct their accumulation to mitochondria—the prime organelle of interest for imaging oxidative stress. Two different targeting strategies were explored. The first involved attachment of a mitochondrial tag (TPP) to the probe structure, employed in the design of **NpFR2**. The second strategy used rhodamine B as a fluorophore with innate mitochondrially-localising abilities. The latter was employed in the development of excitation-ratiometric probes **FRR1** and **FRR2** in which, contrary to the emission ratiometric **FCR1**, flavin was fixed as a FRET-donor and tethered to a rhodamine acceptor. Both these strategies gave rise to unprecedented mitochondrial accumulation of the probes without altering their optical and redox-sensing abilities.

Both **FRR1** and **FRR2** exhibited excellent excitation—ratiometric sensing abilities, the interpretation of the fluorescence output was less straightforward, Nevertheless, the probe design demonstrated that structural variations, primarily at the N-10 position of the flavin scaffold dictate the redox potential of the flavin, a valuable aspect of probe design that would enable modulating the redox potential of future generation flavin-based probes.

Another class of cellular redox-active molecules investigated in this thesis is nicotinamide. The ICT approach, as discussed in Chap. 5, involved conjugating a coumarin molecule to a nicotinamidium ion, and proved successful in achieving two crucial aims of this section—red shifting the photophysical properties and establishing a ratiometric output. However, **NCR3** and **NCR4** underwent irreversible reduction, which was possibly a consequence of its high reduction potential, as suggested by its electrochemical behaviour. Nevertheless, a tertiary pyridinium nitrogen in the nicotinamide ring was identified as an important structural requirement for redox-active behaviour of nicotinamides and needs to be conserved in future probe design. In addition, this cationic structural feature was possibly directing a mitochondrial localisation profile.

A search of the literature reveals that most of the redox sensors developed to date have been applied in imaging biological systems only once when published, highlighting a huge disconnect between the community of chemists who develop these valuable probes and the biological research community that can use these tools to investigate key scientific questions in their domain. While the primary aim of this thesis was to design and develop ratiometric redox sensors a key aspect was to then demonstrate the immense potential of small molecule fluorescent sensors not only in imaging a variety of biological systems using different imaging modalities but also in answering more crucial biological questions.

The developed probes have been investigated in diverse biological systems, as detailed in Chaps. 6–8. The *in cellulo* interrogations revealed that mouse embryonic stem cells with no functional copy of CTR1 ($CTR1^{-/-}$) exhibited low levels of mitochondrial oxidative capacity which could be one of underlying mechanisms hampering the differentiation of these cells. Evaluating the levels of copper in these mutants, either through the use of a fluorescent copper probe or other techniques that quantify metal levels such as GF-AAS or ICP-MS, alongside the redox state imaging of these cells would enable deeper understanding of ESC differentiation processes. PGRMC1 mutations in pancreatic cancer cells revealed that the wild type and triple mutants which exhibited more oxidising mitochondrial matrix demonstrated a less oxidising cytoplasmic environment, such observations could lay the foundation for mechanistic understanding of the progression and consequently the therapeutic approaches towards pancreatic cancer.

Ex vivo studies in isolated mouse hepatocytes established the upper tolerance limits of H_2O_2 treatment and confirmed that deuterated PUFA demonstrated promising antioxidant properties. These preliminary studies can be followed by investigating oxidative stress levels in mice transfected with agents that either compromise or promote functions performed by the liver. Interrogations on the haematopoietic cells communicated the variations in the mitochondrial oxidative capacity not only in different cell types residing in different haematopoietic organs of an adult mouse, but also revealed the changes at different stages of embryogenesis. The information derived from these studies could enable future investigations in new directions. Furthermore, the strength of small molecule redox probes is further evidenced by the preliminary studies performed in non-mammalian systems, such as bacterial biofilms and *C. Elegans*, which establish the applicability of the probes in such systems and provide evidence for the value of further investigations to answer more serious biological questions.

9.1   Future Work

The investigations performed in this work elucidate the design strategies towards the development of ratiometric redox probes based on flavin and nicotinamide scaffold including their utility in a variety of biological systems. However, further modifications could be made, particularly with respect to achieving complete reversibility of redox response in **FCR1**. This could be addressed by performing detailed electrochemical reduction-oxidation mechanistic studies on the probe itself or perhaps by suitably modifying the probe structure to identify the structural component contributing towards this quasi-reversible redox behaviour of **FCR1**.

In addition, derivatives of **FCR1** can be developed by modifying the structural components particularly at C-7, C-8 and N-10 positions on the scaffold that would modulate the redox potential of the flavin moiety and hence the probe itself (Fig. 9.1a, b). This would enable the development of a tool-box of similar probes with different redox-active potential that would find applicability in sensing variations in cellular redox potential, for example, in the progression of prostate cancer.

While FRET has proven to be an extremely valuable strategy for the development of ratiometric probes, it would be valuable to assess the outcomes of employing the first strategy that involved incorporating a fluorophore within the flavin scaffold, by either suitably modifying the naphthalimide structure or using a fluorophore with longer excitation-emission profiles, such as coumarin (Fig. 9.1c). Although flavin-based probes have shown more promise for use as reversible redox sensors, nicotinamide-based probes also possess immense potential and necessitate future investigations involving suitable structural modifications to troubleshoot the irreversible reduction of such probes.

Following the extensive biological interrogations performed with the developed redox sensors certainly merit their use in further research involving disease models to answer biological questions demanding serious attention. The results obtained

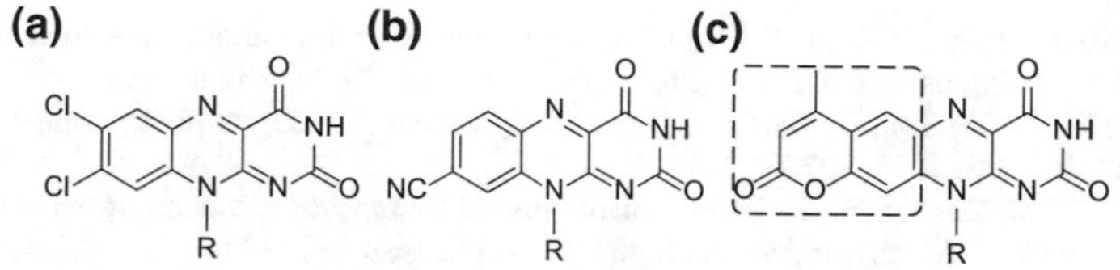

Fig. 9.1 Chemical structures of proposed flavin derivatives illustrating groups on the scaffold that contribute towards tuning the redox potential. **a** 7,8-dichloroisoalloxazine, **b** 8-cyanoisoalloxazine. The R groups can be ethyl or tetraacetylribose. **c** Proposed chemical structural design of a ratiometric probe comprising of a coumarin molecule incorporated within the flavin molecule

from imaging 3-D spheroids can be further understood by using the probe to study spheroids cultured from cells that express the hypoxia responsive element conjugated to a fluorescent protein. This would possibly enable in deconvoluting the probe response in relation to hypoxic tumour micro-environments.

This work has illustrated the design strategies towards the development of ratiometric probes based on naturally-occurring redox-active molecules, flavins and nicotinamides, identifying key structural features required for reversible redox sensing abilities and tuning their redox potential. In addition to desirable photophysical redox responsive properties, the probes exhibited no cytotoxicity under working conditions, did not perturb the redox homoeostasis of the cells and demonstrated promising abilities to report on variations in oxidative capacities in a wide variety of systems. These probes therefore satisfy the criteria of an effective cellular redox probe identified in Sect. 1.7, and certainly warrant further attention and use in studying oxidative stress in biology. Far-reaching consequences of understanding the dynamics of redox state in biology, enabled by the use of such reversible redox probes, include deciphering the variations in cellular oxidative capacity (transient of chronic) that help maintain normal physiological health and that bring about the onset and progression of various diseases.

References

1. L. Yuan, W. Lin, K. Zheng, S. Zhu, FRET-based small-molecule fluorescent probes: rational design and bioimaging applications. Acc. Chem. Res. **46**, 1462–1473 (2013)
2. X. Chen, X. Tian, I. Shin, J. Yoon, Fluorescent and luminescent probes for detection of reactive oxygen and nitrogen species. Chem. Soc. Rev. **40**, 4783–804 (2011)
3. C.C. Winterbourn, The challenges of using fluorescent probes to detect and quantify specific reactive oxygen species in living cells. Biochimica et Biophysica Acta (BBA)—General Subjects **1840**, 730–738 (2014)
4. A. Kaur, J.L. Kolanowski, E.J. New, Reversible fluorescent probes for biological redox states. Angew. Chem. Int. Ed. Engl. **55**, 1602–13 (2016)
5. B. Kalyanaraman, V. Darley-Usmar, K.J. Davies, P.A. Dennery, H.J. Forman, M.B. Grisham, G.E. Mann, K. Moore, L.J. Roberts 2nd, H. Ischiropoulos, Measuring reactive oxygen and nitrogen species with fluorescent probes: challenges and limitations. Free Radic. Biol. Med. **52**, 1–6 (2012)

6. J.L. Kolanowski, A. Kaur, E.J. New, Selective and reversible approaches towards imaging redox signaling using small molecule probes. Antioxid. Redox Signal. (2015)

7. C.D. Hein, X.-M. Liu, D. Wang, Click chemistry, a powerful tool for pharmaceutical sciences. Pharm. Res. **25**, 2216–2230 (2008)

8. J.-F. Lutz, Z. Zarafshani, Efficient construction of therapeutics, bioconjugates, biomaterials and bioactive surfaces using azide-alkyne "click" chemistry. Adv. Drug Deliv. Rev. **60**, 958–70 (2008)

Chapter 10
Experimental Methods

10.1 Instrumentation

^{1}H NMR spectra were obtained at 300 K using Bruker DRX 300 or Bruker Ascend 400 and 500 spectrometers at frequencies of 300 MHz, 400 MHz and 500 MHz, respectively. ^{13}C NMR spectra were recorded on these spectrometers at frequencies of 75, 100, and 125 MHz. Deuterated solvents were obtained from Cambridge Isotope Laboratories were used, and internal references were made using the residual solvent peak. The ^{1}H NMR data are reported as chemical shift (δ) in ppm, and coupling constant (J) in Hz. The ^{13}C NMR data are reported as chemical shift (δ) in ppm downfield shift.

Low resolution mass spectrometry was performed on a Finnigan LCQ quadrupole ion trap mass spectrometer operating in positive ion mode with Electrospray Ionisation (ESI) or Atmospheric Pressure Chemical Ionisation (APCI). High resolution mass spectrometry was performed on a Bruker Apex Qe 7T Fourier transform ion cyclotron resonance mass spectrometer operating in positive ion mode and ESI, using an Apollo II ESI/MALDI dual source.

UV-visible (UV-Vis) absorption measurements were collected on an Agilent Cary 60 UV-Vis Spectrophotometer. Local maxima are reported where appropriate. Fluorescence spectra were recorded using a benchtop Varian Cary Eclipse fluorimeter with quartz cuvettes or a Perkin Elmer Enspire Multimode Plate Reader with flat-bottomed clear and black 96-well plates.

Electrochemical measurements were performed using a BAS 100B/W Electrochemical Analyser using glass carbon working electrode, a platinum auxiliary electrode and a silver/silver chloride reference electrode and using a PGSTAT12 AUTOLAB (Metrohm Autolab B.V., Netherlands) electrochemical analyser. A conventional three-electrode cell configuration was used, consisting of a silver wire quasi reference electrode, a platinum gauze auxiliary electrode and a 3 mm diameter glassy carbon disc working electrode.

Spectro-electrochemical measurements were made using a SEC-C (path length: 1 mm) thin layer quartz glass spectro-electrochemical cell kit (CH Instruments).

© Springer International Publishing AG 2018
A. Kaur, *Fluorescent Tools for Imaging Oxidative Stress in Biology*,
Springer Theses, https://doi.org/10.1007/978-3-319-73405-7_10

This consisted of platinum gauze as the working electrode, platinum wire as the counter and Ag/Ag$^+$ non-aqueous reference electrode. Photoluminescence spectra (5 nm bandpass, 1 nm data interval, PMT voltage: 700 V) were collected with a Cary Eclipse Spectrofluorimeter (Varian, Australia). The platinum gauze surface faced the detector and the side of the cell aligned with the excitation beam.

Confocal images were acquired using an Olympus Fluoview FV1000 microscope and a Leica SP5 II confocal and multi-photon microscope, with either a LUCPLFLN 40X air objective lens (NA = 0.60), UPLSAPO 63× water-immersion (NA = 1.20) or 100× oil-immersion (NA = 1.30) objective lens. Images were collected and processed using Leica Application Suite Advanced Fluorescence Version: 2.8.0 build 7266 viewer software. Image analysis of mean fluorescence intensity was performed using FIJI (National Institutes of Health).

Fluorescence lifetime images were collected on a Leica TCS SP5 MP FLIM system containing a tunable Mai Tai Deep See multi-photon laser with a repetition rate of 80 MHz (Spectra-Physics) connected to a Leica DMI6000B-CS inverted microscope. Samples were illuminated with 820 nm laser and emitted light was collected in the de-scanned internal FLIM detectors over the 420–480 nm and 520–600 range using a HCPLAPO 63× water-immersion (NA = 1.20) objective lens. The data was collected with the aid of the B&H SPCM software and the fluorescence lifetimes were determined using time correlated single photon counting (TCSPC) and analysed with SPC Image software (version 3.1.0.0).

Flow cytometric analyses were performed using BD Biosciences LSRFortessa and BD FACSCan 4-colour flow cytometer. Data was analysed using the FlowJo software package (Treestar, Ashland, OR, USA).

10.2   Methods

10.2.1   Synthetic Methods

All solvents used were laboratory grade and were dried over appropriate drying agents when required. MilliQ water was used to prepare all aqueous solutions. Merck 230–400 mesh Kieselgel 60 was used for silica gel column chromatography and Merck Kieselgel 60 0.25 mm F254 precoated sheets were used for analytical thin layer chromatography. Chemicals were obtained from Sigma-Aldrich (St. Louis, MO), Alfa Aesar and Combi Blocks and used as received. Deuterated solvents were obtained from Cambridge Isotope Laboratories.

10.2.2   Photophysical Studies

Photophysical characterisations of all the probes were performed in HEPES buffer (100 mM, pH 7.4). Three spectra were recorded for each experiment at a scan speed of

120 nm/min and 5 nm slit width, which were then averaged to obtain a smoother curve. Quantum yields (Φ) were calculated using fluorescein ($\Phi = 0.95$) and quinine sulfate ($\Phi = 0.55$) as reference. For each unknown, the standard was selected according to closest excitation wavelengths.

For calculation of quantum yield, five concentrations of the probe and standards were prepared. Quinine sulfate standards were prepared in 0.1 M H_2SO_4 and the fluorescein standards in 0.1 M NaOH. Fluorescence spectra of standard and samples were recorded at same excitation wavelengths. Integrated fluorescence intensities were plotted against the absorbance at excitation wavelength for both the standard and the reduced and oxidised probe. The quantum yield was calculated using the equation:

$$\Phi_X = \Phi_S (D_X / D_S)$$

where Φ is the quantum yield, D is slope, S and X represents the standard and the sample respectively.

10.2.2.1 Redox Titration and Cycling

For all reduction-oxidation experiments, stock solutions of reducing and oxidising agents were freshly prepared immediately before the experiment. The fluorescence spectra of a dilute solution (5–10 μM) of the developed redox probes in HEPES buffer were recorded before and following incremental additions of the reducing agent ($Na_2S_2O_4$ or $NaBH_3CN$). Probes were treated with 10–1000 equivalents of freshly prepared solutions of reducing agent to determine the minimum concentration required for complete reduction of the probe. Re-oxidation of the reduced probe was achieved by incremental additions of oxidising agent (H_2O_2) and determining the time required for complete oxidation of the probe for each addition. Redox cycling experiments involved recording fluorescence spectra following the sequential addition of reducing and oxidising agents and plotting the integrated fluorescence intensity (or ratio, for ratiometric probes) against the number of reduction-oxidation cycles. Volume corrections were obtained from fluorescence measurements of probe solutions to which similar volumes of HEPES buffer was added.

10.2.2.2 pH and Metal Response

100 mM buffer solutions for pH 2–10 were prepared. KCl-HCl buffer was used for pH 2.0, and citrate-phosphate buffers were used for pH 3.0–10.0. Performance of the probes in presence of biologically relevant metal ions (Na^+, K^+, Mg^{2+}, Cu^{2+}, Cu^+, Fe^{2+}, Fe^{3+}, Zn^{2+}, Ca^{2+} and Mn^{2+}) was tested with 100 μM concentration of metal ions in 100 mM HEPES buffer. Stock solutions of metal ions (100 mM) were prepared from their corresponding nitrate salts.

10.2.2.3 Re-oxidation with Biological ROS

Re-oxidation of the reduced probes by different ROS was tested using final concentrations of 100–200 μM of different ROS in 100 mM HEPES buffer. Superoxide ($O_2^{-\bullet}$) was added as solid KO_2. Hydrogen peroxide (H_2O_2) and hypochlorite (OCl^-) were delivered from 30 and 4% aqueous solutions, respectively. Hydroxyl radical ($^\bullet OH$) and tert-butoxy radical ($^\bullet OtBu$) were generated by reaction of 1 mM Fe^{2+} with 200 μM H_2O_2 or 200 μM tert-butyl hydroperoxide (TBHP), respectively.

10.2.3 Electrochemical Studies

At the University of Sydney, solutions were prepared at a concentration of 2–5 mM in acetonitrile containing 0.1 M tetrabutylammonium bromide (TBAB) as a supporting electrolyte. The potentials were referenced to the standard hydrogen electrode (SHE). Solutions were degassed with argon for ten minutes prior to measurement. Cyclic voltammograms were recorded using glass carbon working electrode, a platinum auxiliary electrode and a silver/silver chloride reference electrode.

Electrochemical measurements at La Trobe University in collaboration with Dr Conor F. Hogan, were acquired using a conventional three-electrode cell configuration consisting of a silver wire quasi reference electrode, a platinum gauze auxiliary electrode and a 3 mm diameter glassy carbon disc working electrode. The working electrode was sequentially polished with 0.3 and 0.05 μm alumina slurry on a BUEHLER Microcloth®, rinsed with Milli-Q water and sonicated in acetonitrile for 30 s. Following sonication, the electrode was rinsed in acetonitrile and dried with a stream of nitrogen. Solutions were prepared at a concentration of 2 mM in a solution of freshly distilled acetonitrile containing 0.1 M tetrabutylammonium hexafluorophosphate (TBAPF$_6$) as a supporting electrolyte. All the potentials were referenced to the ferrocene/ferrocenium couple as the internal standard (1 mM). All the electrochemical experiments were performed under a nitrogen atmosphere inside a glove box. Spectro-electrochemistry was performed using a platinum gauze as the working electrode, platinum wire as the counter and Ag/Ag^+ non-aqueous reference electrode. Within the spectrofluorimeter, the spectro-electrochemical cell was placed such that the platinum gauze surface faced the detector and the side of the cell aligned with the excitation beam. An appropriate reduction potential was applied, and a series of luminescence spectra were recorded at 12 s intervals.

10.2.4 Cell Culture

HeLa human cervical adenocarcinoma cells were obtained from Dr Minh Hyunh at the Australian Centre of Microscopy and Microanalysis, DLD-1 human colon carcinoma cells were obtained from Dr Catherine Chen in the Hambley group and

RAW 264.7 murine macrophage cells were obtained from Dr Rachel Pinto at the Department of Infectious Diseases and Immunology, University of Sydney. The cell lines were maintained in exponential growth as monolayers in Advanced DMEM (a basal medium that requires reduced serum supplementation compared to DMEM). For HeLa cells, the growth medium was supplemented with 2% foetal calf serum (FCS), 1% glutamine and 1% antibiotic-antimycotic (AA); for DLD-1 and RAW 264.7 cells, 4% FCS, 1% glutamine and 1% AA were used. The cells were incubated under standard culturing conditions at 37 °C with 5% (v/v) CO_2 under humidified conditions. HeLa and DLD-1 cells were sub-cultured using 0.25% trypsin to facilitate cell dissociation, whereas RAW 264.7 cells were dissociated using a cell scraper.

10.2.4.1 Cytotoxicity Assay

Cytotoxicity assays were performed using the standard MTT (3-(4,5-dimethylthiazol-2-yl)-2,5-diphenyltetrazolium bromide) assay to determine their IC_{50} value [1]. In a 96-well plate, approximately 1×10^4 HeLa or RAW 264.7 cells in 100 μL complete medium (Adv. DMEM supplemented with FCS, glutamine and AA) were seeded into each well and allowed to adhere overnight in an incubator. 1 and 10 mM stock solutions of the compounds of interest were prepared in DMSO. The stock solutions were diluted into the wells containing cells to final concentrations ranging from 0 to 160 μM. Control wells were treated with corresponding volumes of DMSO. Each treatment was performed in triplicates. After a 24 h incubation period, 20 μL of MTT solution (2.5 mg/mL in phosphate-buffered saline (PBS)) was added to each well and the cells were incubated for additional 4 h, allowing the MTT to be converted to a purple formazan product by the mitochondrial dehydrogenase of viable cells. The medium in the wells was then replaced with 150 μL of DMSO and the plates were shaken for 1 min. The absorbance of each well was recorded at 600 nm using a plate reader, wherein the intensity of absorbance correlates to cell viability. The IC_{50} value represents the minimum concentration of the probe necessary for 50% reduction in cell viability. Each MTT assay was performed three times.

10.2.5 Confocal Microscopy

10.2.5.1 Confocal Microscopy of Cell Monolayers

Approximately 5×10^4 HeLa or RAW 264.7 cells in complete medium were plated onto glass bottom dishes (35 mm, coverslip no. 1.5, MatTek Corporation) and allowed to adhere overnight in an incubator. For all treatments with DMSO-based stock solutions were made such that the final concentration of DMSO in cell media was less than 0.1%. Complete medium containing the probe was prepared by diluting probe stock solutions (in DMSO) into 3 mL of complete media to obtain a final concentration of 10–30 μM. The cells were then incubated with the probe containing media for 15–30 min before being washed with PBS and maintained in FluoroBrite™ DMEM media supplemented with 10% foetal calf serum for the duration of imaging. For measurement of the oxidative capacity of cells in different environments, cells were first treated by the addition of 50 μL of the freshly prepared stock solution of reducing agent (N-acetylcysteine, NAC) or oxidising agent (H_2O_2) in PBS to a final concentration of 50 μM in complete media and incubated for 30 min before being washed with PBS. Control cells were treated with 50 μL of PBS to complete media and incubated for 30 min. The cells were then treated with the appropriate concentration of the probe.

For co-localisation studies, HeLa or RAW 264.7 cells were treated with the desired concentration of the probe followed by treatment with the commercially available tracker dyes, Mitotracker DeepRed FM (100 nM, 15 min) or Lysotracker DeepRed FM (100 nM, 15 min) before being washed with PBS and prepared for imaging. The single stain control cells were treated separately with DMSO stock solutions of the probe, Mitotracker DeepRed FM (100 nM, 15 min) or Lysotracker DeepRed FM (100 nM, 15 min).

Confocal images were acquired using an Olympus Fluoview FV1000 microscope or Leica SP5 II confocal and multi-photon microscope. Excitation light of 458 and 488 nm was provided by an argon laser and 633 nm by the HeNe laser. Two photon excitation light of 820 nm was provided by the Mai Tai Deep See Ti:Sapphire femtosecond pulsed laser. Images were analysed using FIJI (National Institutes of Health).

10.2.5.2 Confocal Microscopy of Multicellular Spheroids

96-well plates were coated with 40 μL sterile agarose solution in PBS (0.75% w/v). $1–2.5 \times 10^4$ DLD-1 cells in 100 μL complete medium were seeded into each well and incubated for 3–4 days to allow cell aggregation. The spheroids were then transferred onto glass bottom dishes and dosed with desired concentrations of the probes for 24 h. Confocal microscopy was performed using Leica SP5 II confocal and multi-photon microscope. Image analysis was performed using FIJI (National Institutes of Health).

10.2.5.3 Confocal Microscopy of Mouse Hepatocytes

The isolated hepatocytes were plated at a density of 10,000 cells/well in a 96-well plate in low glucose DMEM supplemented and incubated overnight. Cells were treated with incremental concentrations of H_2O_2 ranging from 100 to 500 μM for 30 min. The cells were then washed and incubated in low glucose DMEM containing 20 μM **NpFR1** for a period of 15 min. This was followed by washing the hepatocytes twice with PBS, and maintaining them in FACS buffer (PBS supplemented with 0.1% FBS) for the duration of imaging. The images were acquired using BD Pathway™ 855, a high-throughput live cell imaging station. Images were acquired using 40X objective upon excitation 405 nm. **NpFR1** emission was recorded using a 450 nm long pass filter, with a higher intensity of fluorescence indicating a higher oxidative capacity. Image analysis was performed using FIJI (National Institute of Health).

To test the antioxidant properties of lipids, 10,000 hepatocytes in each well of a 96 well plate overnight followed by an overnight treatment with palmitate and deuterated linoleic and linolenic acids. Following the chronic lipid treatments, hepatocytes were treated with H_2O_2 (100–400 μM) for 30 min. Hepatocytes were then washed and incubated with 20 μM **NpFR1** for 15 min, washed an imaged in FACS buffer (λ_{ex} = 405 nm and λ_{em} = 450 nm long pass). Image analysis was performed using FIJI (National Institute of Health).

10.2.5.4 Confocal Microscopy of *C. elegans*

The wild-type Bristol strain (N2) of *C. elegans*, were maintained on lawns of *Escherischia coli* OP 50 strain. Prior to treatment with probes the adult worms were filtered using a 40 μm nylon mesh and incubated in different tubes containing 100 μM of each probe in S-basal medium supplemented with *E. coli* OP 50. The worms were incubated at 25 °C or 35 °C for 1 h with constant mixing. Following standard protocols, the worms were then washed and allowed to rest in probe-free S-basal medium for 20 min and washed thrice with S-basal. The worms were then transferred onto glass slides coated with 3% agar pads containing 2% sodium azide. The immobilised worms were then imaged using a Leica SP8 confocal microscope and the images were acquired with a 10× objective. Image analysis was performed using FIJI (National Institutes of Health).

10.2.5.5 Confocal Microscopy of Bacterial Biofilms

Biofilms were prepared on glass slides by Dr Penesyan from Gram-positive (Bacillus cereus-BC) and Gram-negative (Acinetobacter baumannii-AB and Pseudomonas aeruginosa-PA) bacterial strains using continuous flow-cell system over a period of 3–5 days. Prior to imaging the biofilms were maintained in LB medium containing Live/Dead® BacLight stain along with vehicle control (DMSO), **NpFR1** (50 μM) or **FCR1** (20 μM) for 15 min followed by washing with probe-free LB medium for 10 min after which the slides were imaged using the Olympus FV1000 confocal micropscope. Z-stacks were obtained using the 10X objective and were analysed by Dr Penesyan using a Matlab program. Ratio images for **FCR1** were obtained using the RatioPlus plugin for FIJI (National Institutes of Health).

10.2.5.6 Fluorescence Lifetime Imaging Microscopy

Fluorescence lifetime images were collected on a Leica TCS SP5 MP FLIM system containing a tunable Mai Tai Deep See multi-photon laser with a repetition rate of 80 MHz (Spectra-Physics) connected to a Leica DMI6000B-CS inverted microscope.

Samples were illuminated with 820 nm laser and emitted light was collected in the de-scanned internal FLIM detectors over the 420–480 nm and 520–600 range using a HCPLAPO 63× water-immersion (NA = 1.20) objective lens. The data was collected with the aid of the B&H SPCM software and the fluorescence lifetimes were determined using time correlated single photon counting (TCSPC) and analysed with SPC Image software (version 3.1.0.0). The instrument response function was derived from the decay curve of urea. 512 × 512 pixel images were collected and 3X binning applied for analysis to ensure that at least 1×10^4 photons per pixel were analysed. Each sample was scanned for 120 s. Analysis of fluorescence lifetimes in cells required individually fitted mono or bi-exponential curves to obtain average χ^2 values closest to 1. The data ware analysed from 10 different regions of interest from 3 independent experiments. The mean life times were obtained by calculating a decay matrix of each pixel on the 512 × 512 image. The obtained images were pseudo-coloured using a BGR LUT (blue green red look up table) ranging between 0.9 and 2.5 ns.

10.2.6   Flow Cytometry

10.2.6.1   Flow Cytometry of CTR1-knockout Embryonic Stem Cells

Cells were cultured in ES Maintenance Media which comprised of DMEM supplemented with 20% FBS, Glutamax, 10 mL/L sodium pyruvate, 25 mg/L Sodium pyruvate, 39 μL/L MTG, penicillin and streptomycin and 105 units/mL of Leukemia Inhibitory Factor. The cells were grown in a gelatinised 6-well dish at 37 °C and 5% CO_2. For measurement of the oxidative capacity, embryonic stem cells were first treated by the addition of 50 μL of the freshly prepared stock solution of reducing agent (Nacetylcysteine, NAC) or oxidising agent (H_2O_2) in PBS to a final concentration of 50 μM in complete media and incubated for 30 min before being washed with PBS. Control cells were treated with 50 μL of PBS to complete media and incubated for 30 min. The cells were then treated with **NpFR2** (20 μM) for 15 min following which the cells were washed with PBS and detached using TrypLE (350 μL). The cells were then centrifuged, resuspended in FACS buffer (PBS + 0.5% Bovine serum albumin, BSA). Cells were immediately analysed using a BD FACSCan 4-colour flow cytometer. Data obtained were analysed using FlowJo software (Tree Star).

Three murine embryonic stem cell lines were used to analyse the role of CTR1 in differentiation; CTR1$^{+/+}$ cells, CTR1$^{+/-}$ and CTR1$^{-/-}$. These cell lines had been prepared and exposed to mesodermal and ectodermal differentiation conditions by Mr Kurt Brigden at the School of Medicine.

Mesodermal Differentiation

Media conditioned for mesodermal media consists of Iscoves modified dulbeccos medium (IMDM) supplemented with 20% FBS, 25 mg/L ascorbic acid, 39 μL/L MTG and penicillin/streptomycin. Cells were resuspended in 1 mL mesodermal differentiation media and counted using a haemocytometer. 1×10^5 cells from each cell line were seeded out into non-adhesive 23 mm plate in 2 mL of mesodermal differentiation media and allowed to grow for 5 days.

Neuronal Media

12-well plates were gelatinised for 1 hour. Cells were trypsinised 1 day after passage and centrifuged at 1200 rpm for 5 min. Cells were resuspended in ES Maintenance Media and counted using a haemocytometer. Cells were seeded out at a density of 2.5×10^4 for each cell line and made up to a volume of 500 μL using neuronal media; 50% v/v DMEM/F12, 50% neural basal media, 13.4 μL/L MTG, 1% B27 supplement media and 0.5% N2 supplement media. Media was changed day 1 after plating, then every second day thereafter.

The differentiated cells were treated with **NpFR2** (20 μM) for 15 min following which the cells were washed with PBS and detached using TrypLE (350 μL). The cells were then centrifuged, resuspended in FACS buffer (PBS + 0.5% Bovine serum albumin, BSA). Cells were immediately analysed using a BD FACScan 4-colour flow cytometer. Data obtained were analysed using FlowJo software (Tree Star).

10.2.6.2 Flow Cytometry of Haematopoietic Cells

All animal studies were performed in accordance with animal ethical guidelines as approved by the Animal Ethics Committee at the University of Sydney. For flow cytometry experiments with **NpFR2** in mouse haematopoietic cells, adult Quackenbush Swiss male mice were killed by cervical dislocation. Bone marrow, thymus and spleen were immediately dissected. Bone marrow single cell suspensions were prepared by flushing the femora with 5 mL of PBS with a 22G needle and syringe. Thymus and spleen single cell suspensions were prepared by passing the tissues through a 40 μm mesh with the plunger of a 5 mL syringe. All single cell suspensions were then further filtered through 20 μm nylon mesh to remove clumps. Approximately 1×10^6 cells were used per stain. Cells were aliquoted and treated with PBS, DMSO, NAC or H_2O_2 for 30 min at 37 °C, washed and incubated with 20 μM **NpFR2** for 15 min at RT. Cells were then washed and incubated with antibodies recognising surface proteins of live mouse haematopoietic cells for 30 min, washed again with PBS and resuspended in FACS buffer (PBS + 0.5% Bovine serum albumin (BSA) + 1 μM propidium iodide). Cells were immediately analysed using a BD FACSCan 4-colour flow cytometer.

Flow cytometric analyses on cells treated with **FCR2** were performed using BD biosciences LSRFortessa equipped with a 56 mW 405 nm coherent laser. HeLa cells were treated with **FCR2** (10 μM, 15 min) alone; as well as after treatment with reducing (NAC, 50 μM, 30 min) or oxidising agent (H_2O_2, 50 μM, 30 min). The cells were then washed and resuspended in PBS before analysis on the LSRFortessa. Emission intensities were acquired by detectors centered around 450 nm (425–475 nm) and 560 nm (550–570 nm). Approximately 5×10^4 events were collected for each run with appropriate gating applied to isolate healthy and single cells.

For studies of the haematopoietic cell differentiation during mouse embryonic development, timed matings were established as described [2]. Adult bone marrow single cell suspensions were prepared by flushing the femora with 5 mL of PBS with a 22G needle and syringe, spleen single cell suspensions were prepared by passing the tissues through a 40 μm mesh with the plunger of a 5 mL syringe. Foetal blood was obtained by exsanguination of individual conceptuses as described [3]. Foetal liver was dissected and dispersed into a single cell suspension as described previously [4]. Single cell suspensions were then further filtered through 20 μm nylon mesh to remove clumps, centrifuged and resuspended in 1 mL of FACS Buffer (PBS + 0.5% BSA). Cells were aliquoted and incubated with a combination of Ter-119 antibody conjugated to Alexa Fluor 647 (Biolegend, San Diego, CA, USA) and either **FRR1** or **FRR2** (to a final concentration of 20 μM) for 15 min at RT. Cells were then washed with PBS and resuspended in FACS buffer containing propidium iodide (final concentration 1 μg/mL) for dead cell exclusion. Cells were immediately analysed using a BD FACScan 4-colour flow cytometer. Data obtained were analysed using FlowJo software (Tree Star).

10.3   Syntheses

10.3.1   6-Bromo-2-propyl-1H-benzo[de]isoquinoline-1,3(2H)-dione (2)

To a solution of 4-bromo-1,8-naphthalic anhydride (2.99 g, 10.8 mmol) in EtOH (100 mL) was added N-propylamine (0.696 g, 11.8 mmol). The mixture was heated to reflux for 17 h before the solvent was removed under vacuum and the product recrystallised from EtOH to give yellow crystals of **2** (2.47 g, 7.78 mmol, 72%). M.p. 142–144 °C. ESI (m/z): $[M]^+$ calculated for $C_{15}H_{13}{}^{79}BrNO_2$, 317.01; found, 317.12. ^{1}H NMR (400 MHz, CDCl$_3$): δ (ppm) 8.65 (d, J = 8.4, 1H), 8.55 (d, J = 8.5, 1H), 8.40 (d, J = 7.9, 1H), 8.03 (d, J = 7.8, 1H), 7.84 (t, J = 7.6, 1H), 3.72 (t, J = 7.7, 2H), 1.78–1.75 (m, 2H), 1.21 (t, J = 7.5, 3H).

10.3.2 6-Bromo-5-nitro-2-propyl-1H-benzo[de]isoquinoline-1,3(2H)-dione (3)

To a solution of **2** (0.719 g, 2.26 mmol) in concentrated sulfuric acid (20 mL), sodium nitrate (0.92 g, 2.26 mmol) was added and the solution stirred for 30 min at -10 °C and then for 3 h at RT. The mixture was added slowly to ice water (200 mL) and the suspension filtered, washed with water and recrystallised from EtOH to give **3** as a pale yellow solid (0.704 g, 1.94 mmol, 86%). M.p. 147–150 °C. ESI (m/z): $[M]^+$ calculated for $C_{15}H_{11}{}^{79}BrN_2O_4$, 361.99; found, 361.23. ^{1}H NMR (300 MHz, CDCl$_3$): δ (ppm) 8.81 (s, 1H), 8.77 (d, J = 7.3, 2H), 8.01 (t, J = 7.8, 1H), 3.52 (t, J = 7.7, 2H), 1.68–1.66 (m, 2H), 1.15 (t, J = 7.5, 3H).

10.3.3 6-((3-Bromopropyl)amino)-5-nitro-2-propyl-1H-benzo[de]isoquinoline-1,3(2H)-dione (4)

3-Bromopropylamine hydrobromide (1.72 g, 7.85 mmol) was added to a solution of **3** (1.00 g, 2.65 mmol) and DIPEA (1.03 g, 7.85 mmol) in MeCN (50 mL). The mixture was stirred for 2 h at RT under N$_2$ before the solvent was evaporated under vacuum and the residue recrystallised in EtOH to give **4** as a yellow crystalline solid (0.835 g, 1.98 mmol, 75%). M.p. 177–179 °C. ESI (m/z): $[M]^+$ calculated for $C_{18}H_{18}{}^{79}BrN_3O_4$, 419.05; found, 418.18. ^{1}H NMR (400 MHz, CDCl$_3$): δ (ppm) 9.90 (br. s, 1H), 9.24 (s, 1H) 8.65 (m, 2H), 7.68 (t, J = 8.0, 1H), 3.81 (t, J = 7.7, 2H), 3.52 (t, J = 7.7, 2H), 3.32 (t, J = 7.7, 2H), 2.15–2.13 (m, 2H) 1.68–1.65 (m, 2H), 1.12 (t, J = 7.5, 3H).

10.3.4   Attempted Synthesis of 5-Amino-6-((3-bromopropyl) amino)-2-propyl-1H-benzo[de]isoquinoline-1,3(2H)-dione (5)

Stannous chloride dihydrate (0.51 g, 2.2 mmol) was added to a suspension of **4** (0.14 g, 0.34 mmol) and 32% HCl (3 mL) under a N_2 atmosphere. The suspension was heated to reflux for 3 h before being allowed to stir at RT overnight. The mixture was poured onto deionised ice (5 g) and the solution adjusted to pH 11 with 5 M NaOH. The suspension was extracted with DCM (3 × 50 mL) and washed sequentially with water, saturated $NaHCO_3$ solution and brine. The combined organic extracts were dried over Na_2SO_4 and evaporated to dryness. The resulting gel like substance was subjected to silica gel column chromatography (1:1; hexane:EtOAc). However, despite several attempts the desired product could not be isolated.

10.3.5   6-Bromo-2-(3-bromopropyl)-1H-benzo[de]isoquinoline-1,3(2H)-dione (8) [5]

To a solution of 4-bromo-1,8-naphthalic anhydride (2.98 g, 10.8 mmol) in EtOH (100 mL) was added 3-bromopropylamine hydrobromide (2.58 g, 11.8 mmol). The mixture was heated to reflux for 17 h before the solvent was removed under vacuum and the product recrystallised in EtOH to give yellow crystals of **8** (2.44 g, 6.15 mmol, 57%). M.p. 152–154 °C (lit. value 154–155 °C [5]). ESI (m/z): [M+H]$^+$ calculated for $C_{15}H_{12}{}^{79}Br_2NO_2$, 396.91; found, 397.02. ^{1}H NMR (400 MHz, CDCl$_3$): δ (ppm) 8.65 (d, $J = 8.4$, 1H), 8.55 (d, $J = 8.5$, 1H), 8.40 (d, $J = 7.9$, 1H), 8.03 (d, $J = 7.8$, 1H), 7.84 (t, $J = 7.6$, 1H), 3.72 (t, $J = 7.7$, 2H), 3.43 (t, $J = 7.5$, 2H), 2.14–2.12 (m, 2H).

10.3.6 6-Bromo-2-(3-bromopropyl)-5-nitro-1H-benzo[de]isoquinoline-1,3(2H)-dione (9)

To a solution of **8** (0.897 g, 2.26 mmol) in sulfuric acid (20 mL), sodium nitrate (0.192 g, 2.6 mmol) was added and the solution stirred for 30 min at -10 °C and then for 3 h at RT. The mixture was added slowly to ice-water (200 mL) and the suspension filtered, washed with water and recrystallised from EtOH to give **9** as a pale yellow solid (0.786 g, 1.78 mmol, 79%). M.p. 151–152 °C. ESI (m/z): [M+H]$^+$ calculated for C$_{15}$H$_{11}$^{79}Br$_2$N$_2$O$_4$, 441.90; found, 441.86. ^{1}H NMR (300 MHz, CDCl$_3$): δ (ppm) 8.81 (s, 1H), 8.77 (d, J = 7.3, 2H), 8.01 (t, J = 7.8, 1H), 3.52 (t, J = 7.7, 2H), 3.48 (t, J = 7.5, 2H), 2.12–2.09 (m, 2H).

10.3.7 2-(3-Bromopropyl)-5-nitro-6-(propylamino)-1H-benzo[de]isoquinoline-1,3(2H)-dione (10)

N-propylamine (0.471 g, 7.90 mmol) was added to a solution of **9** (0.786 g, 1.78 mmol) in MeCN (30 mL). The mixture was stirred for 1 h at RT under N$_2$ before the solvent was evaporated under vacuum and the residue recrystallised in EtOH to give **10** as a yellow crystalline solid (0.631 g, 1.51 mmol, 83%). M.p. 159–160 °C. ESI (m/z): [M]$^+$ calculated for C$_{18}$H$_{18}$^{79}BrN$_3$O$_4$, 419.05; found, 419.21. ^{1}H NMR (400 MHz, CDCl$_3$): δ (ppm) 9.90 (br. s, 1H), 9.24 (s, 1H) 8.65–8.62 (m, 2H), 7.68 (t, J =

8.0, 1H), 4.13 (t, J = 7.6, 2H), 3.94 (t, J = 7.5, 2H), 3.21 (t, J = 7.5, 2H), 2.07–2.04 (m, 2H), 1.43–1.40 (m, 2H), 0.97 (t, J = 7.6, 3H).

10.3.8   *Attempted Synthesis of (3-(5-nitro-1,3-dioxo-6-(propylamino)-1H-benzo[de]isoquinolin-2(3H)-yl)propyl)triphenylphosphonium (11)*

10 (0.631 g, 1.51 mmol) was added to a solution of triphenylphosphine (0.792 g, 3.02 mmol) in MeCN (10 mL) and heated at reflux conditions under a N_2 atmosphere for 48 h. The reaction mixture was monitored by TLC and mass spectrometry, both of which indicated the presence of unreacted starting material.

10.3.9   *(3-Aminopropyl)triphenylphosphonium bromide hydrobromide (15)*

3-Bromopropylamine hydrobromide (1.00 g, 3.82 mmol) and triphenylphosphine (0.838 g, 3.82 mmol) were added to 5 mL of MeCN and the resulting suspension was heated to reflux for 12 h. The reaction mixture was cooled to RT and hexane (15 mL) was added. The resulting solid was dissolved in isopropanol (100 mL), a minimal amount of diethyl ether (30 mL) was added and the solution left overnight in the refrigerator to give **15** as fine colourless crystals (0.860 g, 1.79 mmol, 47%). M.p. 261–262 °C. ESI (m/z): $[M]^+$ calculated for $C_{21}H_{24}{}^{79}Br_2NP$, 479.00; found, 481.04. ^{1}H NMR (400 MHz, DMSO-d_6): δ (ppm) 7.8–7.75 (m, 15H) 3.74–3.71 (m, 2H), 2.93 (t, J = 7.6, 2H), 1.90–1.87 (m, 2H).

10.3.10 (3-(6-Bromo-1,3-dioxo-1H-benzo[de]isoquinolin-2(3H)-yl)propyl)triphenylphosphonium (16)

To a solution of 4-bromo-1,8-naphthalic anhydride (2.99 g, 10.8 mmol) in EtOH (100 mL), was added **15** (5.67 g, 11.8 mmol). The mixture was heated to reflux for 17 h before the solvent was removed under vacuum and the product recrystallised in EtOAc to give **16** as yellow crystals (4.38 g, 7.56 mmol, 70%). M.p. 245–247 °C. ESI (m/z): [M]$^+$ calculated for $C_{33}H_{26}{}^{79}BrNO_2P$, 578.09; found, 577.94. ^{1}H NMR (400 MHz, DMSO-d$_6$): δ (ppm) 8.62 (d, $J = 8.4$, 1H), 8.56 (d, $J = 8.5$, 1H), 8.40 (d, $J = 7.9$, 1H), 8.35 (d, $J = 7.8$, 1H), 8.04 (t, $J = 7.6$, 1H), 7.5–7.45 (m, 15H), 4.27–4.25 (m, 2H), 3.24–3.21 (m, 2H), 1.88–1.85 (m, 2H).

10.3.11 (3-(6-Bromo-5-nitro-1,3-dioxo-1H-benzo[de]isoquinolin-2(3H)-yl)propyl)triphenylphosphonium (17)

To a solution of **16** (0.73 g, 1.3 mmol) in sulfuric acid (10 mL), sodium nitrate (0.1 g, 1.3 mmol) was added and the resulting solution stirred for 1 h at -10 °C. The mixture was added slowly to ice-water (50 mL), and the resulting suspension filtered, dried and subjected to purification by silica gel column chromatography (2:1; hexane:EtOAc) to give **17** as a pale yellow solid (0.28 g, 0.45 mmol, 36%). M.p. 281–282 °C. ESI (m/z): [M+H]$^+$ calculated for $C_{33}H_{25}BrN_2O_4P$, 623.07; found, 623.26. ^{1}H NMR (400 MHz, DMSO-d$_6$): δ (ppm) 8.64 (d, $J = 8.4$, 1H), 8.59 (d, $J =$

8.5, 1H), 8.45 (d, *J* = 7.9, 1H), 8.11 (t, *J* = 7.6, 1H), 7.5–7.45 (m, 15H), 4.26–4.24 (m, 2H), 3.24–3.22 (m, 2H), 1.87–1.84 (m, 2H).

10.3.12   *(3-(5-Nitro-1,3-dioxo-6-(propylamino)-1H-benzo[de]isoquinolin-2(3H)-yl)propyl)triphenyl phosphonium (18)*

N-Propylamine (0.165 g, 2.80 mmol) was added to a solution of **17** (1.63 g, 2.65 mmol) in MeCN (50 mL). The mixture was stirred for 5 h at RT under N$_2$ before the solvent was evaporated under vacuum. The obtained hygroscopic residue was taken forward in synthesis without further characterisation.

10.3.13   *(3-(5-Amino-1,3-dioxo-6-(propylamino)-1H-benzo[de]isoquinolin-2(3H)-yl)propyl)triphenyl phosphonium (19)*

Stannous chloride dihydrate (0.51 g, 2.2 mmol) was added to a solution of the crude residue (**18**) in 20 mL MeOH and 32% aqueous HCl (3 mL) under a N_2 atmosphere. The suspension was heated to reflux for 1 h. The mixture was poured onto ice prepared form deionised water (5 g) and the solution adjusted to pH 11 with 5 M NaOH. The suspension was extracted with DCM (3 × 50 mL) and washed sequentially with water, saturated NaHCO$_3$ solution and brine. The combined organic extracts were dried over Na$_2$SO$_4$ and evaporated to dryness to give **19** (0.21 g) as a yellow solid which was used in the next step immediately, without further purification to prevent oxidation of the *o*-diamino compound in air.

10.3.14   Triphenyl(3-(4,6,9,11-tetraoxo-13-propyl-9, 10,11,13-tetrahydro-4H-benzo[4,5]isoquinolino [7,6-g]pteridin-5(6H)-yl)propyl)phosphonium (NpFR2)

Alloxan monohydrate (0.11 g, 0.68 mmol) and boric acid (0.45 g, 0.80 mmol) were added to a stirred solution of **19** (0.200 g) in glacial acetic acid (10 mL). The solution was stirred for 7 h under N_2, and then diluted in 100 mL water and filtered. The filtrate was concentrated by evaporating water under vacuum. 200 mL of diethyl ether was added to the crude mixture and sonicated. The suspension was then filtered and the obtained solid was dried and purified by preparative TLC using DCM : MeOH (10:1) as eluent to give **NpFR2** as a bright orange solid (0.13 g, 0.20 mmol, 29%). M.p. 303–305 °C. HRMS: calculated for [M]$^+$ C$_{40}$H$_{33}$N$_5$O$_4$P, 678.22647; found, 678.22652.
^{1}H NMR (500 MHz, DMSO-d$_6$): δ (ppm) 8.91 (d, J = 10.0, 1H, naphthalimide-Ar H), 8.72 (d, J = 10.2, 2H, naphthalimide-Ar H), 8.12 (t, J = 10.0, 1H, flavin-Ar H), 7.77–7.74 (m, 15H, TPP-Ar H), 4.68 (m, 2H, N^8-CH$_2$CH$_2$CH$_2$), 4.25 (t, J = 10.5, 2H, N^{13}-CH$_2$ CH$_2$CH$_3$), 3.77 (t, J = 10.0, 2H, N^8-CH$_2$ CH$_2$ CH$_2$), 2.22–2.19 (m, 2H, N^8-CH$_2$CH$_2$ CH$_2$), 1.86–1.83 (m, 2H, N^{13}-CH$_2$ CH$_2$ CH$_3$), 1.05 (t, J = 10.3, 3H, N^{13}-CH$_2$CH$_2$ CH$_3$). ^{13}C NMR (125 MHz, DMSO-d$_6$): δ (ppm) 160.4, 156.3, 150.7, 139.3, 135.5, 132.7, 132.4 126.5, 116.8, 40.8, 12.4.

10.3.15   (2R,3S,4S)-5-(7,8-dimethyl-2,4-dioxo-3,4-dihydrobenzo[g]pteridin-10(2H)-yl)pentane-1,2,3,4-tetrayl tetraacetate (21)

Riboflavin (10 g, 27 mmol) was stirred in a 1:1 v/v mixture of glacial acetic acid (120 mL) and acetic anhydride (120 mL) for 15 min at 40 °C in the presence of a few drops of 70% perchloric acid. The reaction mixture was then cooled, diluted with water (200 mL) and extracted with dichloromethane (2 × 250 mL). The organic phase was further washed with sodium bicarbonate four times and brine three times. The organic layer was dried over anhydrous Na_2SO_4, filtered, and the solvent removed by rotary evaporation. The orange solid was immediately recrystallised from 220 mL of absolute EtOH. The orange crystals were filtered and washed with cold EtOH to yield **21**, a bright orange crystalline powder (12.5 g, 23.5 mmol, 87%). M.p. 249–251 °C (lit. value 250–252 °C [6]). ^{1}H NMR (200 MHz, CDCl$_3$): δ (ppm) 8.67 (s, 1H), 8.02 (s, 1H), 7.56 (s, 1H), 5.60–5.57 (m, 1H), 5.44–5.41 (m, 3H), 4.43–4.40 (dd, J = 11.0, 2.6, 1H), 4.23 (dd, J = 12.0, 6.4, 1H), 2.56 (s, 3H), 2.44 (s, 3H), 2.28 (s, 3H), 2.21 (s, 3H), 2.01 (s, 3H), 1.78 (s, 3H).

10.3.16   (2R,3S,4S)-5-(3-(2-Hydroxyethyl)-7,8-dimethyl-2,4-dioxo-3,4-dihydrobenzo[g]pteridin-10(2H)-yl)pentane-1,2,3,4-tetrayl tetraacetate (22)

Potassium carbonate (1.5 g, 4.1 mmol) and potassium iodide (0.23 g, 1.3 mmol) were added to a solution of **21** (1.5 g, 2.8 mmol) in DMF (7.5 mL). The suspension was stirred at RT for 15 min under N_2 atmosphere, after which was slowly added a solution of 2-bromoethanol (1.8 g, 14 mmol) in DMF (2 mL). The reaction was stirred at RT under N_2 for 15 h, after which dichloromethane (150 mL) was added and washed sequentially with deionised water, saturated $NaHCO_3$ and brine. The organic layer was dried over anhydrous Na_2SO_4, filtered and the solvent removed by rotary evaporation to yield **22** as an opaque orange solid (1.1 g, 1.8 mmol, 67%). M.p. 210–213 °C. ^{1}H NMR (200 MHz, $CDCl_3$): δ (ppm) 8.01 (s, 1H), 7.76 (s, 1H), 5.60–5.56 (m, 1H), 5.44 (t, J = 5.4, 1H), 5.33 (dd, J = 6.0, 2.8, 1H), 4.99–4.88 (m, 1H), 4.55 (t, J = 4.2, 2H) 4.40 (dd, J = 12, 3.0, 1H), 4.26 (t, J = 4.6, 2H), 4.15 (dd, J = 12, 6.4, 1H), 2.51 (s, 3H), 2.38 (s, 3H), 2.10 (s, 6H), 1.93 (s, 3H), 1.59 (s, 3H).

10.3.17 Attempted Synthesis of (2R,3S,4S)-5-(7,8-Dimethyl-2, 4-dioxo-3-(2-(tosyloxy)ethyl)-3,4-dihydrobenzo[g] pteridin-10(2H)-yl)pentane-1,2,3,4-tetrayl tetraacetate (23)

A solution of **22** (0.50 g, 0.85 mmol) in MeCN (25 mL) was cooled to 0 °C on an icewater-salt bath. Triethylamine (0.13 g, 1.3 mmol) was added and the solution was stirred at 0 °C for 0.5 h, after which a solution of tosyl chloride (0.19 g, 1.0 mmol) in MeCN (15 mL) was added dropwise *via* syringe over a period of 1 h. The solution was stirred as it warmed to RT for 15 h, after which the solvent was removed by rotary evaporation to yield a brown solid. The crude product was purified by flash column chromatography (silica, 50:1 $CHCl_2$:MeOH), but none of the fractions could be identified as the desired product.

10.3.18 6-(Ethyl(phenyl)amino)pyrimidine-2,4(1H,3H)-dione (27)[7]

6-Chlorouracil (0.2 g, 1.3 mmol) and N-ethylaniline (0.5 g, 3.9 mmol) were heated to 170 °C for 20 min with stirring. The reaction mixture was cooled and crushed in ether. The ether layer was decanted and the solid was vigorously stirred in ether:ethanol (5:1), filtered and dried to obtain **27** (0.28 g, 1.19 mmol, 92%) as a colourless solid. M.p. 295–298 °C (lit. value 299 °C [7]). ^{1}H NMR (300 MHz, DMSO-d$_6$): δ (ppm) 10.23 (br. s, 1H), 7.23–7.20 (m, 2H), 6.84–6.81 (m, 3H), 4.44 (q, J = 4.2, 2H), 3.04 (s, 2H), 1.33 (t, J = 4.3, 3H).

10.3.19 10-Ethyl-2,4-dioxo-2,3,4,10-tetrahydrobenzo[g]pteridine 5-Oxide (28) [7]

27 (0.26 g, 1.1 mmol) was dissolved in 5 mL of acetic acid. Sodium nitrite (0.39 g, 5.5 mmol) was added all at once and the reaction mixture was stirred at RT for 3 h, followed by dilution with 10 mL of water. The precipitate obtained was filtered and washed liberally with ice cold water and dried to give **28** as an orange solid (0.25 g, 0.88 mmol, 80%), m.p 301–303 °C (lit. value 305 °C [7]) which was taken forward for reduction without further characterisation.

10.3.20 10-Ethylbenzo[g]pteridine-2,4(3H,10H)-dione (NEF) [7]

To a solution of sodium dithionite (0.54 g, 2.91 mmol) in 10 mL water **28** (0.25 g, 0.97 mmol) was added and the reaction mixture was allowed to stir at RT for 3 h. 2 mL of 30% H_2O_2 was added and the reaction mixture was allowed to stand overnight. The resulting precipitate was filtered, washed with ice cold water and dried to give **NEF** as a yellow solid (0.23 g, 0.81 mmol, 84%). M.p. 341–342 °C (lit. value 347 °C [7]) ^{1}H NMR (300 MHz, DMSO-d$_6$): δ (ppm) 11.37 (br. s, 1H) 8.15 (d, J = 9.0, 1H), 7.98–7.95 (m, J = 9.0, 2H), 7.66–7.63 (m, J = 7.0, 1H), 4.64 (q, J = 4.2, 2H), 1.33 (t, J = 4.3, 3H). ^{13}C NMR (100 MHz, DMSO-d$_6$): δ (ppm) 160.4, 156.3, 150.7, 139.3, 135.5, 132.7, 132.4, 126.5, 116.8, 41.8, 12.4.

10.3.21 10-Ethyl-3-(2-hydroxyethyl)benzo[g]pteridine-2,4(3H,10H)-dione (29)

Potassium carbonate (1.5 g, 4.1 mmol) and potassium iodide (0.23 g, 1.3 mmol) were added to a solution of **NEF** (0.68 g, 2.8 mmol) in DMF (7.5 mL). The suspension was stirred at RT for 15 min under N_2 atmosphere, after which was slowly added a solution of 2-bromoethanol (1.8 g, 14 mmol) in DMF (2 mL). The reaction was stirred at RT under N_2 for 15 h, after which dichloromethane (150 mL) was added and washed sequentially with deionised water, saturated NaHCO$_3$ and brine. The organic layer was dried over anhydrous Na$_2$SO$_4$, filtered and the solvent removed by rotary evaporation to yield **29** as an opaque orange solid (0.41 g, 1.4 mmol, 51%). M.p. 311–312 °C. ^{1}H NMR (300 MHz, DMSO-d$_6$): δ (ppm) 11.37 (br. s, 1H) 8.15 (d, J 9.0, 1H), 7.98–7.95 (m, J = 9.0, 2H), 7.66–7.63 (m, J = 7.0, 1H), 4.64 (q, J = 4.2, 2H), 3.85 (m, 4H), 1.33 (t, J = 4.3, 3H). ^{13}C NMR (100 MHz, DMSO-d$_6$): δ (ppm) 159.8, 155.3, 148.7, 138.6, 133.2, 130.7, 126.4, 124.9, 117.4, 59.4, 46.8, 37.5, 12.1.

10.3.22   Attempted Synthesis of 2-(10-ethyl-2,4-dioxo-4,10-dihydrobenzo[g]pteridin-3(2H)-yl)ethyl 4-methylbenzenesulfonate (30)

A solution of **29** (0.24 g, 0.85 mmol) in MeCN (25 mL) was cooled to 0 °C on an icewater-salt bath. Triethylamine (0.13 g, 1.3 mmol) was added and the solution was stirred at 0 °C for 0.5 h, after which a solution of tosyl chloride (0.19 g, 1.0 mmol) in MeCN (15 mL) was added dropwise *via* syringe over a period of 1 h. The solution was allowed to warm to RT and stirred for 15 h, after which the solvent was removed by rotary evaporation to yield a brown solid. The crude product was purified by flash column chromatography (silica, 50:1 CHCl$_2$:MeOH), but none of the fractions could be identified as the desired product.

10.3.23   2-Oxo-2H-Chromene-3-carboxylate succinimidyl ester (33) [8]

Coumarin-3-carboxylic acid (1.1 g, 6.0 mmol) and *N*-hydroxysuccinimide (0.69 g, 6.0 mmol) were dissolved in 15 mL of anhydrous DMF and stirred at 0 °C for 45 min. *N,N*-dicyclohexylcarbodiimide (1.36 g, 6.6 mmol) was then added and the reaction mixture was stirred for 2 h at RT. The reaction mixture was filtered before 100 mL of isopropanol:hexane (1:20) was added to the filtrate and stirred vigorously. The precipitate obtained was filtered, washed with excess hexane and dried to give **33** as a colourless solid (1.58 g, 5.46 mmol, 91%). M.p. 194–196 °C. ESI (m/z): [M]$^+$ calculated for C$_{14}$H$_9$NO$_6$, 287.04; found, 287.17. ^{1}H NMR (500 MHz, CDCl$_3$): δ (ppm) 8.58 (s, 1H), 7.37 (d, *J* = 9.7, 1H), 6.64 (d, *J* = 5.1, 1H), 6.47 (s, 1H), 2.88 (s, 4H).

10.3.24 N-(3-Bromopropyl)-2-oxo-2H-chromene-3-carboxamide (34)

To **33** (0.499 g, 1.74 mmol) dissolved in dichloromethane, 3-bromopropylamine hydrobromide (0.381 g, 1.74 mmol) and triethylamine (0.41 mL, 3.48 mmol) were added, and the reaction mixture was stirred at RT for 14 h. The reaction mixture was diluted with 50 mL of dichloromethane and sequentially washed with 0.2 M sodium bicarbonate, 0.2 M hydrochloric acid and brine. The organic layer was then dried over Na_2SO_4 and evaporated under reduced pressure to give **34** as a colourless solid (0.477 g, 1.54 mmol, 89%). M.p. 206–208 °C. ESI (m/z): $[M+H]^+$ calculated for $C_{13}H_{12}{}^{79}BrNO_3$, 309.00; found, 309.30. ^1H NMR (500 MHz, CDCl$_3$): δ (ppm) 8.58 (s, 1H), 7.37 (d, J = 9.7, 1H), 6.64 (d, J = 5.0, 1H), 6.47 (s, 1H), 3.64 (t, J = 7.5, 2H), 3.44 (t, J = 7.0, 2H), 2.08–2.05 (m, 2H). ^{13}C NMR (75 MHz, CDCl$_3$): δ (ppm) 161.8, 158.1, 154.7, 129.5, 125.8, 118.7, 114.7, 41.1, 29.8.

10.3.25 N-(3-(10-Ethyl-2,4-dioxo-4,10-dihydrobenzo[g]pteridin-3(2H)-yl)propyl)-2-oxo-2H-chromene-3-carboxamide(FCR)

Potassium carbonate (1.5 g, 4.1 mmol) and potassium iodide (0.23 g, 1.3 mmol) were added to a solution of **NEF** (0.68 g, 2.8 mmol) in DMF (7.5 mL). The suspension was stirred at RT for 15 min under N_2 atmosphere, before the slow addition of a solution of **34** (0.87 g, 2.8 mmol) in DMF (2 mL). The reaction was stirred at RT under N_2 for 15 h, after which chloroform (150 mL) was added and washed sequentially with deionised water, saturated NaHCO$_3$ and brine. The organic layer was dried over anhydrous Na_2SO_4, filtered and the solvent removed by rotary evaporation to yield **FCR** as a yellow solid (0.61 g, 1.28 mmol, 46%). M.p. 173–176 °C. ^1H NMR (500 MHz, CDCl$_3$): δ (ppm) 8.58 (s, 1H), 8.15 (d, J = 9.0, 1H), 7.98–7.95 (m, 2H), 7.66–7.63 (m, 1H), 7.37 (d, J = 9.7, 1H), 6.64 (d, J = 5.2, 1H), 6.47 (s, 1H), 4.64 (q, J = 4.2, 2H), 3.64 (t, J = 7.0, 2H), 3.44 (t, J = 7.3, 2H), 2.08–2.05 (m, 2H), 1.33 (t, J = 4.3, 3H). ^{13}C NMR (75 MHz, CDCl$_3$): δ (ppm) 161.4, 159.7, 158.3, 156.2, 154.4, 149.5, 142.1, 135.7, 126.5, 123.8, 119.4, 116.2, 115.0, 111.3, 46.8, 41.7, 24.9, 17.2.

10.3.26   7-Azido-4-methyl-2H-chromen-2-one (36) [9]

A suspension of 7-amino-4-methyl-2*H*-chromen-2-one (0.50 g, 2.85 mmol) in tetrafluoroboric acid (48 % w/w in H_2O, 1.57 mL) was cooled to −5 °C on an ice bath before a solution of sodium nitrite (0.30 g, 4.3 mmol) in 0.5 mL water was added dropwise. Stirring was continued for 1 h at this temperature. The solid was filtered, washed with ice-cold water and dried under a stream of N_2 gas. The diazonium intermediate was immediately suspended in dry methanol (2.3 mL) at −5 °C before sodium azide (0.20 g, 3.1 mmol) was added and stirred for 1 h at RT. The reaction mixture was concentrated, diluted with water and extracted with ether (3 × 10 mL). The organic phase was dried with Na_2SO_4 and evaporated to give **36** as a yellow solid (0.36 g, 1.79 mmol, 63%). M.p. 161–163 °C. ESI (m/z): $[M]^+$ calculated for $C_{10}H_7N_3O_2$, 201.19; found, 201.08. ^{1}H NMR (CDCl$_3$, 300 MHz):   (ppm) 7.46 (d, *J* = 8.3, 1H), 6.71 (d, *J* = 8.1, 1H), 6.54 (s, 1H), 6.07 (s, 1H), 2.18 (s, 3H).

10.3.27   10-Ethyl-3-(prop-2-yn-1-yl)benzo[g] pteridine-2,4(3H,10H)-dione (37)

Potassium carbonate (1.5 g, 4.1 mmol) and potassium iodide (0.23 g, 1.3 mmol) were added to a solution of **NEF** (0.68 g, 2.8 mmol) in DMF (7.5 mL). The suspension was stirred at RT for 15 min under N_2 atmosphere, after which was slowly added a solution of 3-bromopropyne (0.67 g, 5.6 mmol) in DMF (2 mL). The reaction was stirred at RT under N_2 for 15 h, after which chloroform (150 mL) was added and washed sequentially with deionised water, saturated NaHCO$_3$ and brine. The organic layer was dried over anhydrous Na_2SO_4, filtered and the solvent removed by rotary evaporation to yield **37** (0.42 g, 1.5 mmol, 54%). M.p. 158–161 °C. ^{1}H NMR (300 MHz, DMSO-d$_6$): δ (ppm) 11.37 (br. s, 1H) 8.15 (d, *J* = 9.0, 1H), 7.98–7.95 (m, 2H),

7.66–7.63 (d, J = 7.0, 1H), 4.64 (q, J = 4.2, 2H), 4.22 (s, 2H), 3.08 (s, 1H) 1.33 (t, J = 4.3, 3H). ^{13}C NMR (75 MHz, DMSO-d$_6$): δ (ppm) 162.7, 157.3, 148.4, 135.1, 132.5, 128.3, 126.4, 123.5, 116.7, 78.2, 72.4, 49.2, 11.8.

10.3.28 Attempted Synthesis of 10-Ethyl-3-((1-(4-methyl-2-oxo-2H-chromen-7-yl)-1H-1,2,3-triazol-4-yl)methyl)benzo[g]pteridine-2,4(3H,10H)-dione (38)

37 (0.420 g, 1.5 mmol) and **36** (0.301 g, 1.5 mmol) were dissolved in 50:50 solution of DMF:H$_2$O containing copper iodide (0.09 g, 0.5 mmol) and sodium ascorbate (0.1 g, 0.5 mmol) and the reaction mixture was stirred under N$_2$ for 24 h. TLC indicated the presence of unreacted starting materials.

10.3.29 7-(Diethylamino)-2-oxo-2H-chromene-3-carboxylic acid (40)

4-Diethylaminosalicylaldehyde (3.86 g, 0.02 mol), diethylmalonate (3.2 g, 0.02 mol) and piperidine (2 mL) were combined in absolute ethanol (60 mL) and stirred for 6 h at reflux. Then 10% NaOH (60 mL) solution was added and the mixture was heated under reflux for 15 min. The reaction mixture was cooled to RT and carefully acidified to pH 2 using concentrated hydrochloric acid. The solid obtained was filtered, washed with water, dried and then recrystallised in absolute ethanol to give **40** as bright orange crystals (4.2 g, 0.016 mol, 79%). M.p. 221–223 °C. ESI (m/z): [M+H]$^+$ calculated for C$_{14}$H$_{15}$NO$_4$, 262.10; found, 261.87. ^{1}H NMR (300 MHz, CDCl$_3$): δ (ppm) 8.62 (s, 1H), 7.45–7.42 (m, 1H), 6.79 (d, J = 5.0, 1H), 6.31 (s, 1H), 3.26 (q, J = 5.5, 4H), 1.17 (t, J = 5.0, 6H).

10.3.30   7-(Diethylamino)-2-oxo-2H-chromene-3-carboxylate succinimidyl ester (41)

To a stirring solution of 1-(3-Dimethylaminopropyl)-3-ethylcarbodiimide hydrochloride (0.28 g, 1.5 mmol), and N-hydroxysuccinimide (0.17 g, 1.4 mmol) in anhydrous DMF (10 mL), **40** (0.26 g, 0.98 mmol) dissolved in anhydrous DMF (5 mL) was added dropwise. The reaction was stirred at RT for 48 h in the dark. The resulting yellow mixture was poured into 150 mL of ice/water slurry. The precipitate was then collected by vacuum filtration, washed with 200 mL water, and dried in air overnight to give **41** as a yellow solid (0.319 g, 0.88 mmol, 90%). M.p. 192–195 °C. ESI (m/z): $[M]^+$ calculated for $C_{18}H_{18}N_2O_6$, 358.12; found, 358.20. ^{1}H NMR (500 MHz, CDCl$_3$): δ (ppm) 8.58 (s, 1H), 7.37 (d, J = 9.7, 1H), 6.64 (d, J = 5.0, 1H), 6.47 (s, 1H), 3.40 (q, J = 5.5, 4H), 2.88 (s, 4H), 1.26 (t, J = 5.0, 6H). ^{13}C NMR (100 MHz, CDCl$_3$): δ (ppm) 169.5, 159.3, 159.1, 157.2, 154.3, 151.3, 132.15, 110.3, 107.8, 102.8, 96.9, 45.5, 25.8, 12.6.

10.3.31   N-((Trans-4-aminocyclohexyl)-7-(diethylamino)-2-oxo-2H-chromene-3-carboxamide (42)

A solution of **41** (0.61 g, 1.7 mmol) in dry DMF (50 mL) was added dropwise, to a stirring solution of trans-1,4-diaminocyclohexane (4.0 g, 34 mmol) in dry DMF (10 mL). The reaction was stirred at RT for 24 h in the dark. The mixture was concentrated under vacuum and poured into 150 mL of ice-water slurry to precipitate a yellow solid. The solid was then collected by vacuum filtration, washed with 200 mL water, and dried in air to give **42** as a yellow solid (0.23 g, 1.2 mmol, 74%). ESI (m/z): $[M]^+$ calculated for $C_{20}H_{27}N_3O_3$, 357.21; found, 357.36. M.p. 142–145 °C. ^{1}H NMR (500 MHz, CDCl$_3$): δ (ppm) 8.62 (s, 1H), 7.55 (d, J = 9.0, 1H), 6.82 (dd, J = 9.0, 2.1, 1H), 6.58 (d, J = 2.1, 1H), 3.81–3.78 (m, 1H), 3.52 (q, J = 7.2, 4H), 2.69–2.66 (m, 1H), 1.99 (q, J = 10.1, 4H), 1.37 (m, J = 10.2, 4H), 1.23 (t, J = 6.9, 6H). ^{13}C NMR (125 MHz, CDCl$_3$): δ (ppm) 162.9, 162.4, 157.7, 152.6, 148.2, 131.2, 110.7, 110.0, 108.6, 96.7, 50.1, 48.1, 45.2, 34.6, 31.7, 12.3.

10.3.32 Ethyl-2-(10-Ethyl-2,4-dioxobenzo[g]pteridin-3(2H,4H,10H)-yl)acetate (43)

To a solution of **NEF** (1.0 g, 4.1 mmol) in 50 mL DMF, cesium carbonate (1.6 g, 4.9 mmol) was added and the reaction mixture was stirred under N_2 at 80 °C for 15 min. Ethylbromoacetate (3.45 g, 20.6 mmol) was added dropwise and the reaction mixture was allowed to stir at 80 °C overnight. After dilution with 100 mL of dichloromethane, the reaction mixture was filtered and the solvents were evaporated under reduced pressure. The crude solid was redissolved in 100 mL chloroform and washed with water (3 × 100 mL) and brine. The organic layer was dried over Na_2SO_4 and the solvent was evaporated followed by recrystallisation of crude in ethanol to give **42** as yellow needles (1.05 g, 3.18 mmol, 77%). M.p. 194–196 °C. ^{1}H NMR (500 MHz, DMSO-d_6): δ (ppm) 8.21 (d, J = 9.0, 1H), 8.06–7.95 (m, 2H), 7.70–7.66 (m, 1H), 4.69 (q, J = 4.2, 2H), 4.64 (s, 2H), 4.14 (q, J = 4.2, 2H), 1.35 (t, J = 8.0, 3H), 1.20 (t, J = 8.0, 3H).

10.3.33 2-(10-Ethyl-2,4-dioxobenzo[g] pteridin-3(2H,4H,10H)-yl)acetic acid (44)

43 (1.1 g, 3.1 mmol) was stirred in 15 mL of 32% aqueous hydrochloric acid and stirred at 85 °C for 1 h. The reaction mixture was diluted with excess water and cooled at 4 °CC for about 2–3 h. The precipitate was filtered, dried and recrystallised in 2 M acetic acid to give **44** as fine yellow needles (0.85 g, 2.8 mmol, 93%). M.p. 189–192 °C. ^{1}H NMR (500 MHz, DMSO-d_6): δ (ppm) 8.21 (d, J = 9.0, 1H), 8.06–7.95 (m, 2H), 7.7–7.66 (m, 1H), 4.69 (q, J = 4.2, 2H), 4.64 (s, 2H), 1.35 (t, J = 8.0, 3H).

10.3.34   2-(10-Ethyl-2,4-dioxobenzo[g] pteridin-3(2H,4H,10H)-yl)acetyl chloride (45)

44 (0.2 g, 0.66 mmol) was stirred in 1 mL of neat thionyl chloride for 20 min. Excess thionyl chloride was carefully evaporated under reduced pressure. Residual traces of thionyl chloride were evaporated by subsequent dissolution of crude in toluene followed by evaporation under reduced pressure. The yellow solid was taken forward in synthesis of **FCR1** immediately.

10.3.35   7-(Diethylamino)-N-((1r,4r)-4-(2-(10-ethyl-2,4-dioxo-4,10-dihydrobenzo[g]pteridin-3(2H)-yl)acetamido)cyclohexyl)-2-oxo-2H-chromene-3-carboxamide(FCR1)

To a stirring solution of **45** in 10 mL DMF was added **42** (0.24 g, 0.66 mmol) and DIPEA (0.11 mL, 0.66 mmol). The reaction mixture was stirred for 1 h, then concentrated under reduced pressure, and triturated into cold ether. The obtained solid was filtered and purified by silica gel column chromatography in DCM:MeOH (20:1) to obtain **FCR1** as an orange solid (0.16 g, 0.25 mmol, 38%) M.p. 113–116 °C. APCI-MS: calculated for $[M+H]^+$ $C_{34}H_{37}N_7O_6$, 639.28; found 639.00. ^{1}H NMR (400 MHz, CDCl$_3$): δ (ppm) 8.68 (d, J = 8.0, 1H, coumarin-Ar H), 8.65 (s, 1H, cyclohexyl-NH), 8.32 (d, J = 8.0, 1H, coumarin-Ar H), 7.94–7.91 (m, 1H, coumarin-Ar H), 7.70–7.62 (m, 2H, coumarin-Ar H), 7.42 (d, J = 8.0, 1H, flavin-Ar H), 6.64 (dd, J = 8.0, 1H, flavin-Ar H), 6.48 (d, J = 4.0, 1H, flavin-Ar H), 6.04 (d, J = 4.0, 1H, flavin-Ar H), 4.79–4.76 (m, 4H, flavin-N^{10}-CH$_2$ CH$_3$, N^3-CH$_2$), 3.80–3.95 (br. m, 2H, cyclohexyl-H), 3.45 (q, J = 8.0, 4H, cyclohexyl-H), 2.08–2.05 (m, 4H, cyclohexyl-H), 1.52 (t, J = 8.0, 3H, flavin-N^{10}-CH$_2$ CH$_3$), 1.39 (q, J = 8.0, 4H, coumarin-N-CH$_2$ CH$_3$), 1.24 (t, J = 8.0, 6H, coumarin-N-CH$_2$ CH$_3$). ^{13}C NMR (100 MHz, CDCl$_3$): δ (ppm) 165.9, 162.6, 162.3, 159.5, 157.6, 155.3, 152.5, 148.7, 148,

137.04, 136.2, 135.8, 133.5, 132.3, 131.1, 126.5, 115.1, 110.3, 109.9, 108.4, 96.6, 48.1, 47.8, 45.0, 44.7, 40.4, 31.51,31.51, 31.45, 31.45, 12.4, 12.3.

10.3.36 7-(Diethylamino)-2H-chromen-2-one (47) [10]

4-Diethylaminosalicylaldehyde (1.9 g, 10 mmol), diethylmalonate (3.2 g, 20 mmol) and piperidine (1 mL) were dissolved in absolute ethanol (30 mL) and stirred for 6 h under reflux conditions. Ethanol was evaporated under reduced pressure, and then concentrated HCl (50 mL) and glacial acetic acid (20 mL) were added to the reaction, then heated to reflux for an additional 24 h. The solution was cooled to RT and poured into 100 mL ice water. NaOH solution (40% w/v) was added dropwise to modulate pH of the solution to 5. The pale yellow precipitate was filtered, washed with water, dried, and recrystallised with toluene to give **47** as a yellow solid (1.74 g, 8.0 mmol, 80%). M.p. 183–185 °C. ESI (m/z): [M+H]$^+$ calculated for $C_{13}H_{16}NO_2$, 218.17; found, 218.10. ^{1}H NMR (500 MHz, CDCl$_3$): δ (ppm) 7.53 (d, J = 10.2, 1H), 7.24 (d, J = 9.0, 1H), 6.56 (d, J = 9.0, 1H), 6.51 (s, 1H), 6.06 (d, J = 8.5, 1H), 3.40 (q, J = 5.0, 4H), 1.21 (t, J = 5.3, 6H). ^{13}C NMR (125 MHz, CDCl$_3$): δ (ppm) 162.4, 156.9, 150.8, 143.8, 128.91, 109.3, 108.8, 108.4, 97.7, 44.9, 12.5.

10.3.37 7-(Diethylamino)-2-oxo-2H-chromene-3-carbaldehyde (48) [11]

Anhydrous DMF (2 mL) was added dropwise to POCl$_3$ (2 mL) at 20–50 °C under a N$_2$ atmosphere and stirred for 30 min to afford a red solution. This solution was combined with **47** (1.50 g, 6.91 mmol) and dissolved in 10 mL DMF to afford a scarlet suspension. The mixture was stirred at 60 °C for 12 h and then poured into 100 mL of ice water. NaOH solution (20% w/v) was added to adjust the pH to 7. The crude product was filtered, thoroughly washed with water, dried and recrystallised in absolute ethanol to give **48** as a bright yellow solid (1.20 g, 4.89 mmol, 71%). M.p. 149–151 °C (lit. value 152–154 °C [11]). ESI (m/z): [M+H]$^+$ calculated for $C_{14}H_{16}NO_3$, 246.19; found, 245.96. ^{1}H NMR (400 MHz, CDCl$_3$): δ (ppm) 10.16 (s,

1H), 8.28 (s, 1H), 7.44 (d, J = 8.0, 1H), 6.65 (dd, J = 8.8, 2.4, 1H), 6.52 (d, J = 2.4, 1H), 3.50 (q, J = 8.0, 4H), 1.28 (t, J = 8.0, 6H). ^{13}C NMR (100 MHz, CDCl$_3$): δ (ppm) 187.9, 161.8, 158.9, 153.4, 145.34, 132.5, 114.4, 110.2, 108.3, 97.3, 45.3, 12.4.

10.3.38   Methyl 5-methylnicotinate (50 a) [12]

5-Methylnicotinic acid (0.51 g, 3.75 mmol) and 1 mL of concentrated H$_2$SO$_4$ were added to 50 mL of methanol, and the resulting solution was heated under reflux for 3 h. The reaction mixture was allowed to cool to RT and was concentrated under vacuum. The resultant mixture was extracted with DCM, and the combined organic layers were dried using Na$_2$SO$_4$, filtered, and concentrated to give **50 a** as a viscous oil which crystallised on standing (0.34 g, 2.2 mmol, 60%). M.p. 44–46 °C (lit. value 47–48 °C [12]). ^{1}H NMR (300 MHz, CDCl$_3$): δ (ppm) 9.18 (d, J = 2.5, 1H), 8.34 (d, J = 2.3, 1H), 7.64–7.61 (m, 1H), 3.91 (s, 3H), 2.24 (s, 3H).

10.3.39   Methyl 6-methylnicotinate (50 b) [12]

6-Methylnicotinic acid (0.51 g, 3.75 mmol) and 1 mL of concentrated H$_2$SO$_4$ were added to 50 mL of methanol, and the resulting solution was heated at reflux for 3 h. The reaction mixture was allowed to cool and was concentrated under vacuum. The resultant mixture was extracted with DCM, and the combined organic layers were dried using Na$_2$SO$_4$, filtered, and concentrated to give **50 b** as a viscous oil which crystallised on standing (0.44 g, 2.9 mmol, 77%). M.p. 33–34 °C (lit. value 32–33 °C [12]). ^{1}H NMR (300 MHz, CDCl$_3$): δ (ppm) 9.21 (d, J = 2.5, 1H), 8.42 (dd, J = 9.5, 2.5, 1H), 7.37 (d, J = 9.3, 1H), 3.87 (s, 3H), 2.58 (s, 3H).

10.3.40 5-Methylnicotinamide (51 a) [12]

50 a (1.4 g, 9.3 mmol) was stirred in 28% ammonium hydroxide solution (100 mL) for 6 h at RT. The reaction mixture was concentrated under vacuum, and the residue was recrystallised from ethanol to give **51 a** as a white solid (0.90 g, 6.6 mmol 72%). ^{1}H NMR (300 MHz, CDCl$_3$): δ (ppm) 8.92 (d, J = 2.3, 1H), 8.09 (d, J = 2.5, 1H), 7.34–7.31 (m, 1H), 2.24 (s, 3H).

10.3.41 6-Methylnicotinamide (51 b) [12]

50 b (1.1 g, 7.3 mmol) was stirred in 28% ammonium hydroxide solution (100 mL) for 6 h at RT. The reaction mixture was concentrated under vacuum, and the residue was recrystallised from ethanol to give **51 b** as a white solid (0.85 g, 6.2 mmol, 86%). ^{1}H NMR (300 MHz, CDCl$_3$): δ (ppm) 9.17 (d, J = 2.5, 1H), 8.52 (dd, J = 9.0, 2.5, 1H), 7.54 (d, J = 9.0, 1H), 2.27 (s, 3H).

10.3.42 6-(2-(7-(Diethylamino)-2-oxo-2H-chromen-3-yl)vinyl)nicotinamide (NCC)

Under N$_2$ atmosphere, **48** (1.0 g, 4.1 mmol) was added to a solution of **51 b** (0.67 g, 4.9 mmol) and p-toluenesulfonic acid (1.9 g, 11 mmol) in anhydrous DMF (30 mL). The resulting solution was heated at 60 °C over night. The crude product was purified by chromatography on silica gel DCM:acetone (90:10) to give **NCC** as a red solid (0.44 g, 1.2 mmol, 30%). M.p. 172–175 °C. ESI (m/z): [M]$^+$ calculated for $C_{21}H_{21}N_3O_3$, 363.16; found, 363.01. ^{1}H NMR (400 MHz, DMSO-d$_6$): δ (ppm) 8.98 (s, J = 5.0, 1H), 8.19 (s, J = 8.0, 1H), 8.15 (d, J = 8.0, 1H), 8.09 (d, J = 8.0, 1H), 7.61 (d, J = 7.0, 1H), 7.56 (s, 2H), 7.52 (m, 1H), 6.64–6.61 (m, 1H), 6.49 (d, J = 3.0, 1H), 3.56 (q, J = 7.2, 4H), 1.18 (t, J = 7.2, 6H).

10.3.43   5-Carbamoyl-1-ethyl-2-methylpyridin -1-ium Iodide (52)

50 b (0.75 g, 5.5 mmol) was stirred in neat ethyliodide at 60 °C over night. The reaction mixture was allowed to cool and the slurry was added to cold acetone. The resulting solid was filtered and washed with acetone:ethanol (75:25) to give **52** as a light yellow solid (0.72, 2.5 mmol) in 45 % yield. ESI (m/z): [M]$^+$ calculated for $C_9H_{13}N_2O$ 165.10; found, 165.47. ^{1}H NMR (300 MHz, CDCl$_3$): δ (ppm) 9.47 (d, J = 9.0, 1H), 9.08–9.06 (m, 1H), 8.73 (d, J = 9.1, 1H), 4.76 (q, J = 7.0, 2H), 2.97 (s, 3H), 1.71 (t, J = 7.2, 3H).

10.3.44   5-Carbamoyl-2-(2-(7-(diethylamino)-2 -oxo-2H-chromen-3-yl)vinyl)-1-ethylpyridin -1-ium (NCR3)

Under N$_2$ atmosphere, **48** (1.0 g, 4.1 mmol) was added to a solution of **52** (1.4 g, 4.9 mmol) and p-toluenesulfonic acid (1.9 g, 11 mmol) in anhydrous DMF (30 mL). The resulting solution was heated at 60 °C over night. The crude product was purified by chromatography on silica gel DCM:acetone (90:10) to give **NCR3** as a deep red solid (0.83 g, 2.2 mmol, 52%). ESI (m/z): [M]$^+$ calculated for $C_{23}H_{26}N_3O_3$, 392.20; found, 392.13. ^{1}H NMR (400 MHz, DMSO-d$_6$): δ (ppm) 9.28 (d, J = 8.0, 1H, nicotinamide-Ar H), 8.72 (d, J = 8.2, 1H, nicotinamide-Ar H), 8.54 (d, J = 8.2,

1H, nicotinamide-Ar H), 8.39 (s, 1H, nicotinamide-NH), 8.36 (s, 1H, nicotinamide-NH), 8.03 (s, 1H, coumarin-Ar H),7.96–7.85 (m, 2H, coumarin-Ar H), 7.57 (d, J = 8.0, 1H, coumarin-Ar H), 6.81 (d, J = 8.0, 1H, vinyl H), 6.62 (s, 1H, vinyl H), 4.71 (q, J = 7.0, 2H, nicotinamide-N- $\underline{CH_2}$ CH_3), 3.50 (q, J = 7.2, 4H, coumarin-N-$\underline{CH_2}$ CH_3), 1.52 (t, J = 7.2, 3H, nicotinamide-N-CH$_2$ $\underline{CH_3}$), 1.14 (t, J = 7.1, 6H, coumarin-N-CH$_2$ $\underline{CH_3}$). ^{13}C NMR (100 MHz, DMSO-d$_6$): δ (ppm) 163.6, 160.4, 157.2, 154.1, 153.1, 147.3, 145.9, 141.9, 141.4, 131.8, 130.6, 125.0, 115.7, 113.7, 111, 109.1, 96.9, 54.4, 45.1, 15.6, 13.0.

10.3.45 *5-Carbamoyl-1-hexyl-2-methylpyridin-1-ium Iodide (53)*

50 b (0.75 g, 5.5 mmol) was stirred in neat hexyliodide at 60 °C over night. The reaction mixture was allowed to cool and the slurry was added to cold acetone. The resulting solid was filtered and washed with acetone:ethanol (75:25) to give **53** as a light yellow solid (0.96, 2.73 mmol, 49%). ESI (m/z): [M]$^+$ calculated for C$_{13}$H$_{21}$N$_2$O, 221.16; found, 221.28. ^{1}H NMR (300 MHz, CDCl$_3$): δ (ppm) 9.32 (d, J = 8.0, 1H), 9.04–9.01 (m, 1H), 8.62 (d, J = 8.0, 1H), 5.18 (t, J = 5.0, 2H), 2.87 (s, 3H), 2.62–2.59 (m, 2H), 1.35–1.32 (m, 6H), 1.07 (t, J = 5.3, 3H).

10.3.46 *5-Carbamoyl-2-(2-(7-(diethylamino)-2-oxo -2H-chromen-3-yl)vinyl) -1-hexylpyridin-1-ium (NCR4)*

Under N_2 atmosphere, **48** (1.0 g, 4.1 mmol) was added to a solution of **53** (1.27 g, 3.64 mmol) and p-toluenesulfonic acid (1.9 g, 11 mmol) in anhydrous DMF (30 mL). The resulting solution was heated at 60 °C over night. The crude product was purified by chromatography on silica gel DCM:acetone (90:10) to give **NCR4** (1.06 g, 1.9 mmol, 48%). ESI (m/z): $[M]^+$ calculated for $C_{27}H_{34}N_3O_3$, 448.26; found, 448.39. ^1H NMR (400 MHz, DMSO-d_6): δ (ppm) 9.29 (s, 1H, nicotinamide-Ar H), 8.74 (d, $J = 8.2$, 1H, nicotinamide-Ar H), 8.56 (d, $J = 8.2$, 1H, nicotinamide-Ar H), 8.42 (s, 1H, nicotinamide-NH), 8.31 (s, 1H, nicotinamide-NH), 8.05 (s, 1H, coumarin-Ar H), 7.93 (d, $J = 8.0$, 2H, coumarin-Ar H), 7.58 (d, $J = 7.0$, 1H, coumarin-Ar H), 6.83 (d, $J = 9.0$, 1H, vinyl H), 6.63 (d, $J = 8.0$, 1H, vinyl H), 4.66 (t, $J = 5.0$, 2H, nicotinamide-N^1-CH$_2$ (CH$_2$)$_4$CH$_3$), 3.51 (q, $J = 7.2$, 4H, coumarin-N-CH$_2$ CH$_3$), 1.87 (quin, $J = 7.2$, 2H, nicotinamide-N^1-CH$_2$ CH$_2$ (CH$_2$)$_3$CH$_3$), 1.41–1.36 (m, 2H, nicotinamide-N^1-(CH$_2$)$_2$ CH$_2$ (CH$_2$)$_2$CH$_3$), 1.26–1.33 (m, 4H, nicotinamide-N^1-(CH$_2$)$_3$ (CH$_2$)$_2$ CH$_3$), 1.15 (t, $J = 5.3$, 6H, coumarin-N-CH$_2$ CH$_3$) 0.85 (t, $J = 7.1$, 3H, nicotinamide-N^1-(CH$_2$)$_5$ CH$_3$). ^{13}C NMR (100 MHz, DMSO-d_6): δ (ppm) 163.6, 160.3, 157.2, 154.2, 153.1, 147.7, 146.2, 141.8, 141.6, 131.8, 130.4, 124.9, 115.9, 113.7, 111.0, 109.1, 96.9, 58.7, 45.1, 31.1, 29.8, 25.8, 22.5, 14.4, 14.4, 13.0, 13.0.

References

1. T. Mosmann, Rapid colorimetric assay for cellular growth and survival: application to proliferation and cytotoxicity assays. J. Immunol. Methods **65**, 55–63 (1983)
2. V.I. Antas, K.W. Brigden, A.J. Prudence, S.T. Fraser, Gastrokine-2 is transiently expressed in the endodermal and endothelial cells of the maturing mouse yolk sac. Gene Expr. Patterns **16**, 69–74 (2014)
3. S.T. Fraser, M. Ogawa, T. Yokomizo, Y. Ito, S. Nishikawa, S.-I. Nishikawa, Putative intermediate precursor between hematogenic endothelial cells and blood cells in the developing embryo. Dev. Growth Differ. **45**, 63–75 (2003)
4. J. Isern, S.T. Fraser, Z. He, M.H. Baron, The fetal liver is a niche for maturation of primitive erythroid cells. Proc. Natl. Acad. Sci. U S A **105**, 6662–6667 (2008)
5. Y.-Y. Zhang, J.-L. Mi, C.-H. Zhou, X.-D. Zhou, Synthesis of novel fluconazoliums and their evaluation for antibacterial and antifungal activities. Eur. J. Med. Chem. **46**, 4391–4402 (2011)
6. J.-H. Jeon, T. Kakuta, K. Tanaka, Y. Chujo, Facile design of organicinorganic hybrid gels for molecular recognition of nucleoside triphosphates. Bioorgan. Med. Chem. Lett. **25**, 2050–2055 (2015)
7. F. Yoneda, Y. Sakuma, M. Ichiba, K. Shinomura, Syntheses of isoalloxazines and isoalloxazine 5-oxides. A new synthesis of riboflavin. J. Am. Chem. Soc. **98**, 830–835 (1976)
8. L.-S. Zhou, K.-W. Yang, L. Feng, J.-M. Xiao, C.-C. Liu, Y.-L. Zhang, M.W. Crowder, Novel fluorescent risedronates: synthesis, photodynamic inactivation and imaging of Bacillus subtilis. Bioorgan. Med. Chem. Lett. **23**, 949–954 (2013)
9. S. Ast, P.J. Rutledge, M.H. Todd, Reversing the triazole topology in a cyclam-triazole-dye ligand gives a 10-fold brighter signal response to Zn2+ in aqueous solution. Eur. J. Inorg. Chem. **2012**, 5611–5615 (2012)
10. T.-K. Kim, D.-N. Lee, H.-J. Kim, Highly selective fluorescent sensor for homocysteine and cysteine. Tetrahedron. Lett. **49**, 4879–4881 (2008)

11. L. Long, X. Li, D. Zhang, S. Meng, J. Zhang, X. Sun, C. Zhang, L. Zhou, L. Wang, Amino-coumarin based fluorescence ratiometric sensors for acidic pH and their application for living cells imaging. RSC Adv. **3**, 12204–12209 (2013)
12. J.I. Seeman, H.V. Secor, C.G. Chavdarian, E.B. Sanders, R.L. Bassfield, J.F. Whidby, Steric and conformational effects in nicotine chemistry. J. Organ. Chem. **46**, 3040–3048 (1981)

Appendix A
NMR Spectra

See Appendix Figs. A.1, A.2, A.3, A.4, A.5, A.6, A.7, A.8, A.9 and A.10.

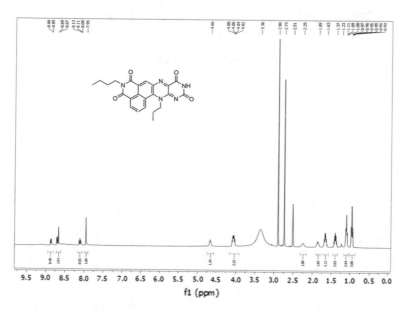

Fig. A.1 ^{1}H NMR spectrum of **NpFR1** (500 MHz, DMSO-d$_6$)

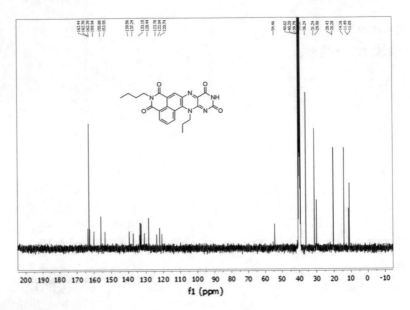

Fig. A.2 ^{13}C NMR spectrum of **NpFR1** (125 MHz, DMSO-d$_6$)

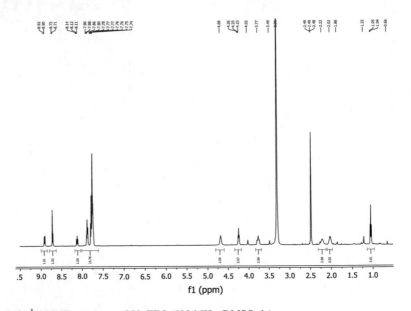

Fig. A.3 ^{1}H NMR spectrum of **NpFR2** (500 MHz, DMSO-d$_6$)

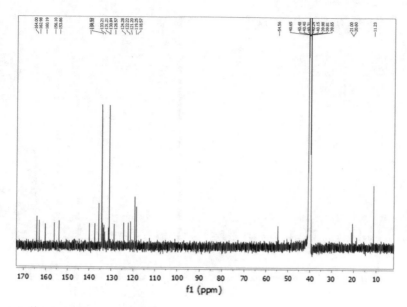

Fig. A.4 ^{13}C NMR spectrum of **NpFR2** (125 MHz, DMSO-d$_6$)

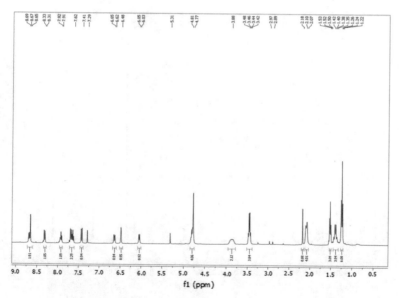

Fig. A.5 ^{1}H NMR spectrum of **FCR1** (500 MHz, DMSO-d$_6$)

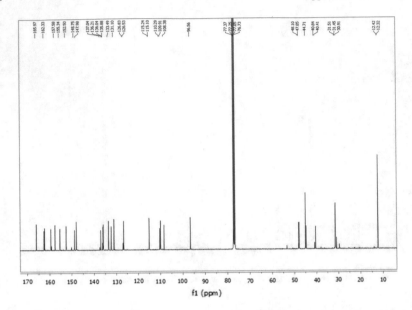

Fig. A.6 ^{13}C NMR spectrum of **FCR1** (125 MHz, DMSO-d$_6$)

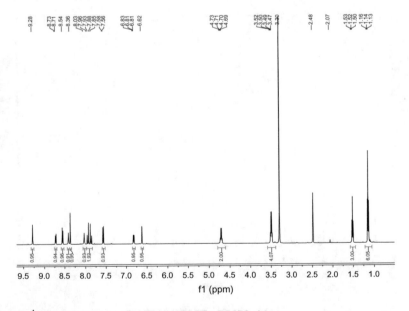

Fig. A.7 ^{1}H NMR spectrum of **NCR3** (400 MHz, DMSO-d$_6$)

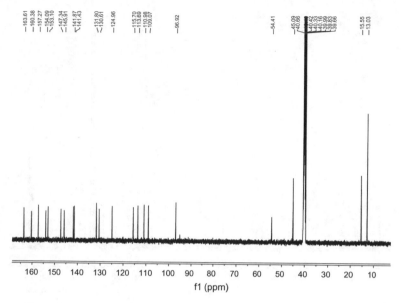

Fig. A.8 ^{13}C NMR spectrum of **NCR3** (100 MHz, DMSO-d$_6$)

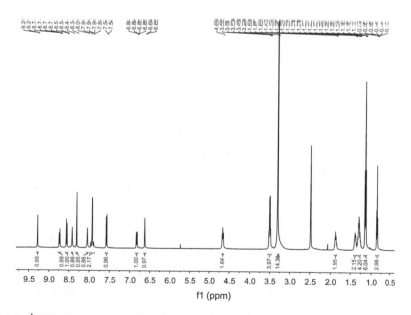

Fig. A.9 ^{1}H NMR spectrum of **NCR4** (500 MHz, DMSO-d$_6$)

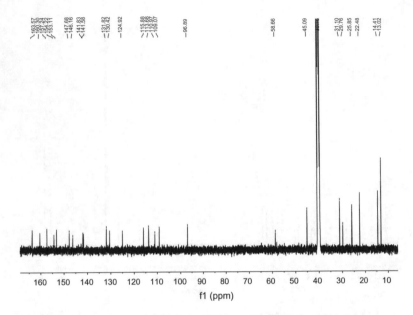

Fig. A.10 ^{13}C NMR spectrum of **NCR4** (125 MHz, DMSO-d$_6$)

Printed in the United States
By Bookmasters